GraphQL Best Practices

Gain hands-on experience with schema design, security, and error handling

Artur Czemiel

‹packt›

GraphQL Best Practices

Copyright © 2024 Packt Publishing

Group Product Manager: Kaustubh Manglurkar

Publishing Product Manager: Urvi Sambhav Shah

Book Project Manager: Sonam Pandey

Senior Editor: Hayden Edwards

Technical Editor: Simran Ali

Copy Editor: Safis Editing

Indexer: Subalakshmi Govindhan

Production Designer: Alishon Mendonca

DevRel Marketing Coordinators: Anamika Singh and Nivedita Pandey

First published: October 2024

Production reference: 1120924

Published by Packt Publishing Ltd.

Grosvenor House

11 St Paul's Square

Birmingham

B3 1RB, UK.

ISBN: 978-1-83546-714-5

www.packtpub.com

To my wife, Paulina, and my kids, Aleksander and Wiktor, for always being there.

– Artur Czemiel

Contributors

About the author

Artur Czemiel is a co-founder of the GraphQL Editor and owner of the Aexol agency, with over 15 years of experience in GraphQL, TypeScript, and full stack development. He is also the creator of the GraphQL client generator, GraphQL Zeus (available on GitHub), and the GraphQL backend engine, Axolotl.

Before this, Artur was a visual effects (VFX) specialist, writing Python scripts for movie effects, and patented an algorithm for weighing small objects with a smartphone.

Artur holds a computer science engineering degree, with his thesis on the measurement of temperature using a smartphone and sound waves.

I want to thank the people who have been close to me and supported me, especially my wife, Paulina; the co-founders of my company, Tomek and Kamil; my friends, Wojtek and Maciek; my friend Sarin, who invited me to the conference to talk about GraphQL; my parents, Leszek and Jola (who is a writer, so she helped me a little); my mother-in-law, Ewa, who helps me raise my kids; and my siblings, Magda, Kuba, Mikołaj, and Maks, who always support me.

I also want to thank the Packt team for being super professional.

About the reviewers

Fadi William Ghali Abdelmessih is a seasoned full stack software engineer with extensive experience in both frontend and backend development, specializing in React, Spring Boot, and Node.js. His expertise also extends to mobile application development for both iOS and Android, spanning the banking and entertainment industries, where he has developed robust applications tailored to these sectors. At Microsoft, he served as a partner technical consultant, focusing on cloud application development.

Fadi holds a master's degree in computer fundamentals and engineering from Polytech Nice Sophia, France, and a BSc in computer engineering from the French University in Egypt.

Martyn Robotham is an experienced web developer with nearly 10 years of experience developing full stack and frontend web applications. He has also been a mentor to others, guiding them and helping them grow in their careers. He has worked with a variety of technologies in .NET and frontend web development, working across a diverse set of technologies and industries.

Ryan Dsouza is a full stack cloud-native engineer, architecting apps using optimal practices from development to deployment. He has worked with Node, React, Python, Docker, and various AWS services such as Lambda, ECS, SQS, SNS, and AppSync. He also has expertise in CI/CD solutions and advocates for infrastructure as code and serverless solutions.

Table of Contents

Part 2 - Schema-First Design Patterns

3

Crafting Effective GraphQL Schemas   51

4

Building Pipes   73

Part 3 - Exploring Possible Ways to Use GraphQL

5

Transitioning from REST to GraphQL   91

Part 4 - Advanced GraphQL

8

Executing Schema-First Systems   149

9

Working on the Frontend with GraphQL   179

10

Keeping Data Secure   219

Part 5 - From an Idea to a Working Project

14

From an Idea to a Working Project – Backend Development with GraphQL and TypeScript   283

15

From an Idea to a Working Project – Frontend Integration with GraphQL and TypeScript   335

Preface

Specifying the structure of web backends has always been an integral part of most software projects. However, it can often be a challenge, due to the numerous technologies available for specification. As a result, the software industry is constantly searching for the ideal solution.

I have been interested in specifying software products from the very beginning of my career when I was a VFX developer. I developed plugins for a popular 3D software, and I frequently received requests to create small web-based apps that would enhance a company's workflow. I would ask myself the question, "*Wouldn't it be sufficient to just specify the data structure that holds business data and generate everything else out of it?*"

In this regard, GraphQL emerges as a highly promising candidate. It offers a compelling alternative that addresses the complexities faced by developers, providing a clear and concise approach to backend specification – not only because it is a good language to specify structures and systems, but also because it's already widely adopted and developers have already accepted it.

The goal of this book is for you to learn GraphQL and full stack development based on the GraphQL schema. Within this book, you will learn how to create transformer libraries, use AI to generate GraphQL schema and docs, build a frontend communication layer for GraphQL, and build great schema-first backends.

Who this book is for

This book is for developers interested in advancing their knowledge of GraphQL, from beginners curious about the basics to experienced developers looking to refine their schema design and security strategies. It's particularly ideal for backend developers, full stack professionals, and frontend developers who wish to understand GraphQL from the ground up and apply advanced techniques in real-world scenarios. Additionally, this book is a must-read for anyone looking to make their applications more efficient and robust using GraphQL.

What this book covers

In *Chapter 1, Unveiling the Holy Grail of Communication – GraphQL*, you will learn what GraphQL is and how it changes the way development teams communicate.

In *Chapter 2, Applying an AI-Friendly Approach to GraphQL*, you will learn how to generate parts of GraphQL using AI.

In *Chapter 3, Crafting Effective GraphQL Schemas*, we will examine all the features of GraphQL and learn how to create consumer-friendly schemas.

In *Chapter 4, Building Pipes*, you will learn how to use access-based resolvers in GraphQL.

In *Chapter 5, Transitioning from REST to GraphQL*, you will learn how to move from REST to GraphQL and write GraphQL proxies for existing REST backends.

In *Chapter 6, Defining GraphQL Transformers*, you will learn what a transformer is and how to write one of your own.

In *Chapter 7, Understanding GraphQL Federation*, we will try to understand what federation is and why it was created.

In *Chapter 8, Executing Schema-First Systems*, we will create a GraphQL-based backend for a Questions and Answers project.

In *Chapter 9, Working on the Frontend with GraphQL*, we will create a frontend for our Questions and Answers project with a GraphQL layer.

In *Chapter 10, Keeping Data Secure*, you will learn how to design GraphQL schema with security in mind and avoid exposing sensitive data.

In *Chapter 11, Describing Errors in GraphQL*, you will learn how GraphQL can help shape the data of returned errors.

In *Chapter 12, Documenting your Schema*, you will learn how to write good documentation for your GraphQL-based systems.

In *Chapter 13, Tackling Schemas with Visualization*, you will learn how to read visual graphs and filter out specific information.

In *Chapter 14, From an Idea to a Working Project – Backend Development with GraphQL and TypeScript*, we will create a complicated backend for a booking system that a small business such as a hairdresser could use.

In *Chapter 15, From an Idea to a Working Project – Frontend Integration with GraphQL and TypeScript*, we will continue our booking system project by creating a frontend for the system.

To get the most out of this book

To get the most out of this book, you just need a computer that runs Node.js and TypeScript with either Windows, macOS, or Linux. You have to be a software developer to write full stack code, but if you're not, the theoretical part of the book will still be useful for you to understand how data is shaped using GraphQL.

Software/hardware covered in the book	Operating system requirements
GraphQL Editor	Windows, macOS, or Linux
TypeScript 5.x	
Axolotl	
GraphQL Yoga	
GraphQL Zeus	
GraphQL Demeter	
React 18.x	
OpenAI	
Shadcn UI	

All the software used throughout the book has open source versions.

After reading this book, deploy at least one full stack project to cement your knowledge.

If you are using the digital version of this book, we advise you to type the code yourself or access the code from the book's GitHub repository (a link is available in the next section). Doing so will help you avoid any potential errors related to the copying and pasting of code.

Download the example code files

You can download the example code files for this book from GitHub at `https://github.com/PacktPublishing/GraphQL-Best-Practices`. If there's an update to the code, it will be updated in the GitHub repository.

We also have other code bundles from our rich catalog of books and videos available at `https://github.com/PacktPublishing/`. Check them out!

Conventions used

There are a number of text conventions used throughout this book.

`Code in text`: Indicates code words in text, database table names, folder names, filenames, file extensions, pathnames, dummy URLs, user input, and Twitter handles. Here is an example: "Within the `__schema` field, we retrieve the `queryType`, `mutationType`, and `subscriptionType` fields."

A block of code is set as follows:

```
type Todo {
  id: ID!
  title: String!
  completed: Boolean!
}
```

Bold: Indicates a new term, an important word, or words that you see on screen. For instance, words in menus or dialog boxes appear in **bold**. Here is an example: "To do this, click the **Filter by root type** button at the top menu of GraphQL Editor."

> **Tips or important notes**
> Appear like this.

Get in touch

Feedback from our readers is always welcome.

General feedback: If you have questions about any aspect of this book, email us at `customercare@packtpub.com` and mention the book title in the subject of your message.

Errata: Although we have taken every care to ensure the accuracy of our content, mistakes do happen. If you have found a mistake in this book, we would be grateful if you would report this to us. Please visit `www.packtpub.com/support/errata` and fill in the form.

Piracy: If you come across any illegal copies of our works in any form on the internet, we would be grateful if you would provide us with the location address or website name. Please contact us at `copyright@packt.com` with a link to the material.

If you are interested in becoming an author: If there is a topic that you have expertise in and you are interested in either writing or contributing to a book, please visit `authors.packtpub.com`.

Share Your Thoughts

Once you've read *GraphQL Best Practices*, we'd love to hear your thoughts! Scan the QR code below to go straight to the Amazon review page for this book and share your feedback.

https://packt.link/r/1-835-46714-8

Your review is important to us and the tech community and will help us make sure we're delivering excellent quality content.

Download a free PDF copy of this book

Thanks for purchasing this book!

Do you like to read on the go but are unable to carry your print books everywhere?

Is your eBook purchase not compatible with the device of your choice?

Don't worry, now with every Packt book you get a DRM-free PDF version of that book at no cost.

Read anywhere, any place, on any device. Search, copy, and paste code from your favorite technical books directly into your application.

The perks don't stop there, you can get exclusive access to discounts, newsletters, and great free content in your inbox daily

Follow these simple steps to get the benefits:

1. Scan the QR code or visit the link below

https://packt.link/free-ebook/978-1-83546-714-5

2. Submit your proof of purchase
3. That's it! We'll send your free PDF and other benefits to your email directly

Part 1 -
Why GraphQL?

In the first part of the book, you will learn what GraphQL is and why it is a game-changer for development team communications. You will also learn how we can utilize AI to help us build consistent schemas and generate documentation.

This part contains the following chapters:

- *Chapter 1, Unveiling the Holy Grail of Communication – GraphQL*
- *Chapter 2, Applying an AI-Friendly Approach to GraphQL*

1

Unveiling the Holy Grail of Communication – GraphQL

GraphQL seemed to appear both recently and also a long time ago. We live in such times that it's hard to define. We can definitely consider the year 2015 as the beginning of GraphQL, as that was when Facebook decided it was the right time to make GraphQL open source and available to everyone. Although GraphQL was primarily created to limit data transfer, as a side effect, it also reduced the need for extensive communication within the development team.

I became deeply interested in GraphQL technology around 2018. Since 2016, I had been trying to solve the communication problem between frontend and backend teams by building a TypeScript-based system that generated a REST application skeleton for backend developers and "enforced" them to implement a system with the appropriate types for each endpoint and return exactly what was specified by the system designer (who worked in a visual graph editor). In turn, frontend developers received a generated library with exactly the same input and output types as an SDK. I was elated when Tomasz and Piotr Karwatka introduced me to GraphQL and said, "*Just build it on top of GraphQL because it's hyped and it's good.*" That's when my GraphQL journey truly began.

But what is **GraphQL**? While most programming languages are languages that execute code, fulfill requirements, and run our software, GraphQL is a language of definition, documentation, and specification. But most importantly, it is a language of ideas. This language is our "Platonic cave," and our software should be a shadow of each idea created in this cave.

Whether you have already used GraphQL or want to take the first step in learning GraphQL, this book is for you. I can't promise it will be easy, but what you will achieve after reading it is something no one can take away from you.

In this chapter, we will learn how to distinguish between the GraphQL query language and the GraphQL Schema Definition Language. We will see why GraphQL is so important in today's world, not only allowing us to connect with businesspeople but also showing us how to communicate better within a team through a concise system definition.

We will also learn why GraphQL exists and how it actually impacts the speed of creating new projects. We will focus on these advantages of GraphQL that facilitate communication within a development team. With this knowledge, you will surely understand and be able to explain to others how to significantly improve communication within a development team, and you will never want to go back to just using REST.

So, in this chapter, we'll cover the following topics:

- Understanding the advantage of built-in introspection
- Working with GraphQL as a source of truth
- Creating self-documenting schemas
- Understanding the GraphQL query language
- Mocking GraphQL responses

Technical requirements

To follow this chapter and run the mock backend example, you will need the following:

- Node.js with a version higher than 18
- GraphQL Demeter: `https://github.com/graphql-editor/graphql-demeter`
- VS Code, GraphQL Editor, Sublime Text, or any text editor to edit GraphQL files to play with the mock backend

A GitHub repository has been created for this book, and you can find the code for this chapter here: `https://github.com/PacktPublishing/GraphQL-Best-Practices/tree/main/chapter-01`.

Understanding the advantage of built-in introspection

Why does GraphQL bring about such significant changes? This is primarily because of **introspection**. In practice, this means that frontend developers, mobile developers, or any API consumers can send a query to the backend asking about the available resolvers, their parameters, and what will be returned from the backend. Thanks to this mechanism, developers can efficiently communicate with the API without diving into the system documentation.

Introspection is a mechanism that allows developers to see the entire structure of a system. Through introspection, we can not only discover what queries we can execute against the backend but also distinguish between queries for data retrieval and mutations for making changes to the system. Additionally, we receive detailed information about the returned types.

Why wasn't this concept invented earlier, considering it is so simple and obvious? Well, it has been invented before, for example, in the form of the OpenAPI Specification. However, the problem lies in the fact that with REST APIs, new versions can be released without necessarily updating the documentation or specification. GraphQL, on the other hand, is designed in such a way that an undocumented type or field does not exist.

But how do we get there? How can we make querying our system so easy? Well, before we can do that, we first need to define what is in our system. To do that, we need to familiarize ourselves with the schema definition language and leverage all its advantages to shape our data structure.

The **Schema Definition Language** (**SDL**) is the foundation of GraphQL because, before allowing queries to be executed, we use the SDL to define their structure, the parameters they accept, and the types they return. Although we will learn more about the schema-first approach in this book, where we define the schema first and then match the code to it, it's worth noting that many systems generate SDL from models created in another programming language (the code-first approach).

The SDL is very minimalist and understandable by anyone who understands data types. It was intentionally kept simple and not overly complicated so that the number of people who understand GraphQL would ultimately be much greater than the number of developers who have worked with it. It's even nice to visualize GraphQL and show it to non-technical people. I guarantee that most of them, based solely on the visualization, will be able to grasp the ideas behind the system.

Here, we have a schema for creating tasks – we can retrieve a list of tasks with their ID, title, and completed status, plus we can create, edit, or delete a task:

```
type Todo {
  id: ID!
  title: String!
  completed: Boolean!
}

type Query {
  todoList: [Todo!]!
}

type Mutation {
  createTodo(title: String!): Todo!
  updateTodo(
    id: ID!,
    title: String,
    completed: Boolean
    ): Todo!
  deleteTodo(id: ID!): ID!
}
```

```
schema {
  query: Query
  mutation: Mutation
}
```

Now, let's imagine that we are a frontend developer who does not have access to the repository or to this file. How can we learn about the backend structure and how to make queries?

Through introspection, of course. Thanks to this benevolent mechanism, the entire team using the API knows exactly what responses will be returned by the backend.

Before we dive into the intricacies of introspection, let's use the built-in GraphiQL tool to communicate with our backend. Tools such as GraphiQL or GraphQL Language Server perform introspection under the hood and, based on that, they create an **abstract syntax tree** (**AST**) that provides us with features such as autocomplete, error checking, and more. When working on our task example, every time we start writing in GraphiQL, we receive autosuggestions. Those suggestions look like this:

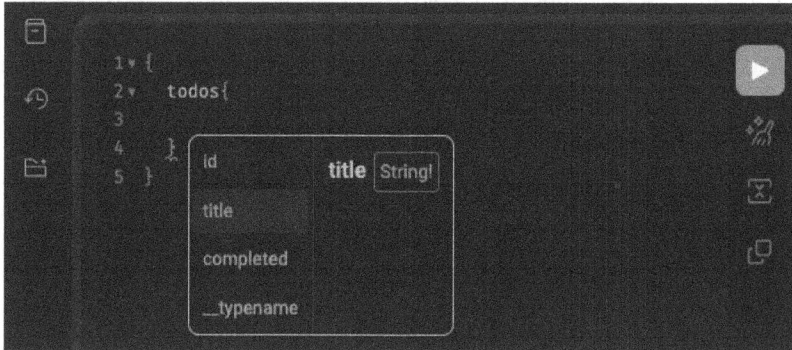

Figure 1.1: GraphiQL query editor

As seen in the figure, after pressing *Ctrl* + the *spacebar*, the GraphiQL interface suggests how we should query the backend thanks to introspection. Think about it like you would think about the code library – every GraphQL schema contains all the information needed to provide this kind of autocomplete system. (I dare say that even a programmer who is not familiar with GraphQL should be able to write queries as well by themselves.)

Now, when we look at our schema again, we can see that it is really basic. It's quite evident that there is a lack of documentation – types and their fields don't have any description – but from a GraphQL perspective, we can still gain insights into its structure. Thanks to the power of GraphQL introspection, we have fields and types documented, enabling tools such as GraphiQL to provide autocomplete suggestions as we write queries:

```
query IntrospectionQuery {
  __schema {
```

```
    queryType { name }
    mutationType { name }
    subscriptionType { name }
    types {
      ...FullType
    }
    directives {
      name
      description
      locations
      args {
        ...InputValue
      }
    }
  }
}
```

The GraphQL introspection query allows us to retrieve the schema of a GraphQL server, providing valuable information about the available types, fields, and directives. Let's break down the query step-by-step:

1. We start with the `query` keyword to indicate that this is a query operation.

2. We name the query `IntrospectionQuery` – you don't have to do this, but to keep good standards, we should name the queries.

3. Inside the query, we access the `__schema` field. This field represents the root of the schema and contains all the information we need.

4. Within the `__schema` field, we retrieve the `queryType`, `mutationType`, and `subscriptionType` fields. These fields represent the root query, mutation, and subscription types, respectively.

5. Next, we access the `types` field, which returns a list of all the types defined in the schema. We use the spread operator (. . .) followed by the `FullType` fragment (an optimization for both the transmitted query and the written query) to include detailed information about each type. You can check the code of the fragment inside the GitHub repository's `chapter-01/example-02-introspection.gql` file. We will also look at fragments again later in the chapter.

6. Finally, we retrieve the `directives` field. Directives allow us to modify the execution of a GraphQL query and provide additional instructions to the server. We retrieve the `name`, `description`, `locations`, and `args` fields for each directive. The `args` field uses the `InputValue` fragment to include detailed information about each argument.

By executing this introspection query, we can gain a comprehensive understanding of the GraphQL schema and its available types, fields, and directives. This information is particularly useful for tools such as GraphQL clients, IDEs, and documentation generators, as it allows them to provide intelligent autocompletion, validation, and documentation assistance to developers working with the schema.

Though it's not very feasible to include the full introspection, query, and explanation in this book, even with a limited representation, it's clear that the introspection showcases how information is gathered from the schema recursively, delving down the tree to about seven levels. This provides a valuable understanding of the schema's hierarchical structure.

As you can see, the schema's structure is important, but if we are unable to access this information, the functionality of GraphQL is hindered. It is the most important advantage over REST. In the upcoming section, we will delve into how GraphQL can serve as a reliable source of truth and explore how to effectively work with it.

Working with GraphQL as a source of truth

I often reminisce about the days without GraphQL with an ironic smile on my face. I remember how, as a frontend developer, I had to call backend developers and ask why a certain field suddenly disappeared or what the name of an endpoint was because I couldn't remember it anymore. Of course, even in REST, there are specification systems that allow for a detailed description of our backend; however, they don't enforce it on the backend developer – they only enable it.

I was already interested in the concept of a single source of truth even before GraphQL, but as soon as I understood the power of GraphQL, I immediately decided to teach it to everyone in our company and turn it into a competitive advantage.

Understanding the source of truth

Let's imagine a scenario where we're starting a project from scratch without any existing systems in place. Typically, the process involves creating wireframes and mockups, and then moving on to the design. Once the design is ready, we begin working on the backend and, later on, the frontend. However, with GraphQL, we have the flexibility to take a different approach. We can start directly from our source of truth, which is GraphQL itself.

From this single source of truth, different teams can consume and discuss this source of truth, enabling them to collaboratively find the best solutions for our system. This approach allows for a more efficient and streamlined development process.

Having a single source of truth also has other benefits. If we have a team member with extensive business knowledge about the solution, we can teach them GraphQL and have them build the schema. This way, we can be sure that the team of developers, designers, QA, and so on will perfectly reflect the project requirements.

Furthermore, we can treat GraphQL as a low-code solution in this scenario. Why? A well-defined schema is excellent nourishment for developers to create proper code. It's much easier to program when you have a well-built foundation. In the future, or perhaps even at the time you are reading this book, there may already be a solution that allows the implementation of a backend system based solely on the schema.

On the other hand, if you are a frontend developer or working on a mobile application, a well-coordinated single source of truth will allow you to write code without waiting for your backend team colleagues. After all, you are capable of generating excellent mocks.

Example project – book library

In this section and for the rest of the chapter, we will explore how to initiate a greenfield project using GraphQL as the source of truth.

> **Note**
>
> A **greenfield project** is a term used to describe a project that starts from scratch in an empty field. The term originates from the construction industry, where a greenfield project is a construction project on an undeveloped site.

Our project will be based around a university library, providing a simple and efficient way for librarians to manage books. They can easily retrieve a list of available books, mark books as borrowed, and update the availability status when books are returned. This schema can be further expanded to incorporate more features, such as keeping track of due dates or generating reports on book lending activity.

By designing the GraphQL schema, we will see how GraphQL will help each member of the team and how a well-defined schema will facilitate everyone's work and reduce communication.

Okay, let's get started. Take a close look at the library schema of the library. It's a simple project, but you will see how much can be determined using GraphQL even before writing the first line of code, or even before the designer draws the first stroke:

```
type Book {
  id: ID!
  title: String!
  author: String!
}

type Query {
  getAllBooks: [Book!]!
```

```
    getBookById(id: ID!): Book
    getAvailableBooks: [Book!]!
}

type Mutation {
  addBook(
    title: String!,
    author: String!,
    genre: String!,
    publicationYear: Int!
    ): Book!
  updateBook(
    id: ID!,
    title: String,
    author: String,
    genre: String,
    publicationYear: Int
    ): Book!
  deleteBook(id: ID!): Book!
  lendBook(
    id: ID!,
    borrower: String!, dueDate: String!
    ): Book!
  returnBook(id: ID!): Book!
}

schema {
  query: Query
  mutation: Mutation
}
```

Let's focus on the most important type, which is Book. Currently, the only information we have about the books are title and author; however, in a library, we need to have much more information about a book than that.

As we shape the schema, we can add the publication date, genre, availability, borrower name, and due date, like so:

```
type Book {
  id: ID!
  title: String!
```

```
    author: String!
    genre: String!
    publicationYear: Int!
    isAvailable: Boolean!
    borrower: String
    dueDate: String
}
```

But just including the author's name in the Book type is not enough. So, let's make the Author type so that we can establish a relationship between the author and the book (though remember that an author can write multiple books, and a book can have multiple authors):

```
type Book{
    id: ID!
    title: String!
    author: [Author!]!
    genre: String!
    publicationYear: Int!
    isAvailable: Boolean!
    borrower: String
    dueDate: String
}
type Author{
    firstName: String!
    lastName: String!
    books: [Book!]!
    photoUrl: String
}
```

Now that we have these two types ready – Book and Author – based on this information, the designer should know what mock values to put on the design when it comes to the detailed view of a book or an author. They should also know what fields will be available and what structure of data to expect.

Furthermore, the designer may suggest that it would be useful to have a book cover and an author photo. Let's add those to our schema and then see what the whole thing looks like:

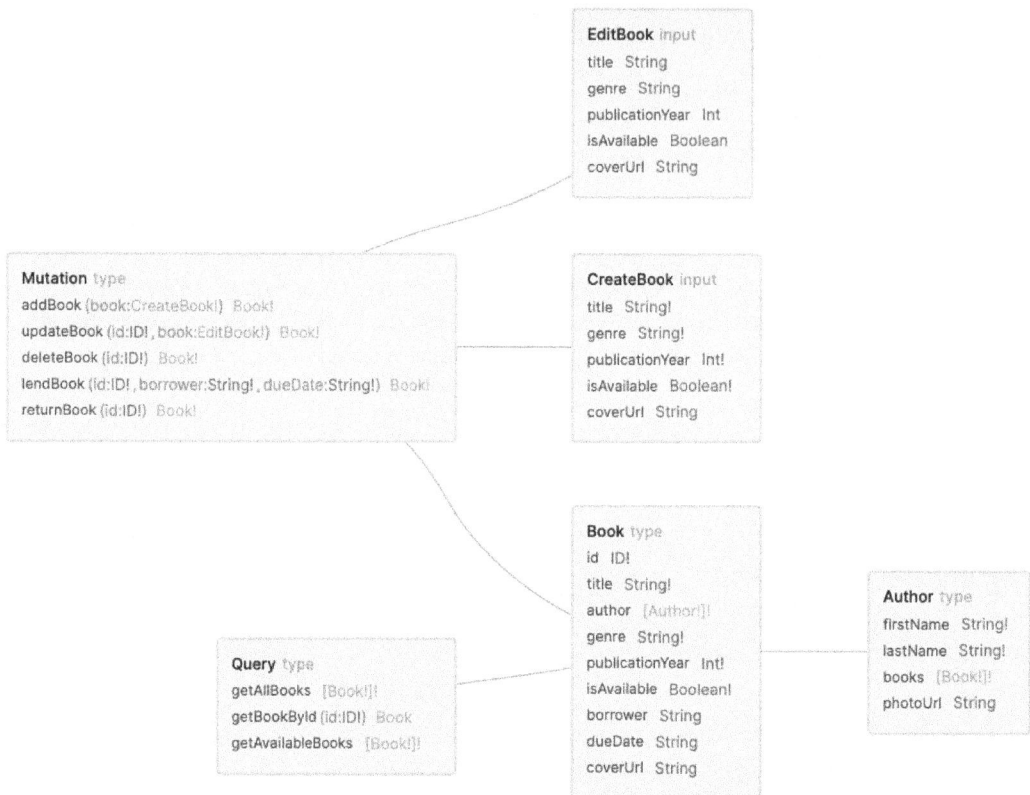

Figure 1.2: Visual schema graph of the library

Do you see it? You don't even need to turn on your computer to see what this system does as a whole. You can comfortably sit down and imagine the exact functionalities of the entire system.

Now, looking at this figure, every backend developer, even those who are not familiar with GraphQL technology, can see the relationships between types and how our system works. On the other hand, designers and frontend developers can get an idea of the views that will be needed in the application. This schema is our source of truth, something that every software project can start with.

> **Note**
>
> GraphQL schemas have become a standard for us when estimating the time and cost of non-standard IT solutions, making it easier to present our understanding of the business logic to clients or project creators. These schemas are not super complicated for them to understand, compared to using something like **Unified Modeling Language (UML)**, especially considering we can also provide additional visualization.

With that, let's move on to the next section, where we will learn about self-documenting schemas.

Creating self-documenting schemas

As you can see from our library schema, even though it doesn't have documentation, just by looking at it, we know exactly what the system based on this schema is used for. This is clear to every member of the team.

Now, let's go further, looking at **self-documenting schemas** – these are GraphQL schemas that contain comprehensive and descriptive naming and connections between nodes, allowing developers to easily understand and interact with the available data and operations. Here, we will look at the different elements that make up these schemas.

Custom scalars

Scalars represent simple primitives in GraphQL. There are five built-in scalars in GraphQL: `String`, `Int`, `Float`, `ID`, and `Boolean`. Out of these five scalars, only `ID` needs to be translated, as it is used to indicate a unique key and can be either a string, a number, or a custom object.

With custom scalars, the situation is different as they behave according to how we define them. While `JSON`, for example, is a scalar that immediately tells you what should be contained within it, `DateTime`, on the other hand, requires documentation on the format of the date and time it represents.

Let's see how the documentation for a `DateTime` scalar should be done:

```
"""
Scalar that holds the date and time.
"""
scalar DateTime
```

Here, we have almost no information about that scalar. Instead, what we need to add is the format of the `DateTime` object and an example, like so:

```
"""
Scalar that holds the date and time.
The most common ISO Date Time Format
yyyy-MM-dd'T'HH:mm:ss. SSSXXX
— for example, "2000-10-31T01:30:00.000-05:00".
"""
scalar DateTime
```

Now, the information about our scalar is clear. Knowing that the date comes in this format from the backend and should be sent in this format from the frontend means that the frontend developer can correctly implement it.

Directives

Directives often require more attention than scalars. Initially, directives were mainly used to provide additional information during query execution, but later, many developers started using them for various schema transformations.

> **Note**
> Built-in directives will be discussed later in this chapter.

Here, I will show you an example of a very popular directive, `@model`, which is used in generative systems to create database models corresponding to GraphQL types:

```
"""
Directives are used to create database models out of GraphQL types.
They provide relationship support as well as handle primaryKey and
unique fields.
"""
directive @model(
  """
  The name of the database table for this model
  """
  tableName: String!
  """
  The primary key field for this model
  """
  primaryKey: String!
  """
  The fields of this model that are unique
  """
  unique: [String!]
  """
  The relationships of this model
  """
  relationships: [Relationship!]
) on OBJECT

input Relationship{
  """
  The type of the related model
  """
  type: String!
  list: Boolean!
  """
```

```
    The foreign key field for this relationship
    """
    foreignKey: String!
}
```

This directive is well documented now. Of course, we could add some database details and information about what the schema will look like after the generation, but when it comes to documentation, that's it!

> **Note**
>
> This section may have reminded you about transformers, but you'll have to wait until *Chapter 6* to learn more – if you don't know, basically, it's the directive that is used to generate one schema from another. For example, we can use the @model directive on type to specify it is a database model that needs **Create, Read, Update, and Delete (CRUD)** mutations generated out of it.

Fields

Fields can be considered the most important part of the GraphQL syntax, as it is through fields that the executed operation tree continues to grow. Fields have varying levels of importance, and out of respect for this hierarchy, they must be documented.

The most important fields are, of course, Query, Mutation, and Subscription. These fields serve as the main entry points to our system, which is where our attention should be focused the most. Of course, if we use pipes (which will be discussed later in *Chapter 4*), these important fields may also appear a bit deeper in the hierarchy.

We need to describe these fields as best as we can (which doesn't mean the description should be long) so that the developer can embrace our system from the beginning and not get lost in it. So, let's go back to our library example and document our queries and mutations:

```
type Query{
    """
    Get all books that are in the library, including borrowed
      books
    """
    getAllBooks: [Book!]!
    """
    Get a book by its database id
    """
    getBookById(
      id: ID!
    ): Book
    """
    Get all books that are physically in the library.
```

```
    """
    getAvailableBooks: [Book!]!
}
```

This is a much more comfortable situation than schema without documentation itself. By adding descriptions to our fields, we know clearly what we can get from the system.

Enums

Enums are a crucial part of GraphQL schema. They are primitive types like scalars, but they allow only a set of values defined within the SDL. It's no surprise that enums are documented as well. Here, I will show you an example of an enum that may not have any documentation, as well as an example where each enum field definitely needs documentation.

So, here's a self-explanatory enum:

```
enum Locale{
    EN
    FR
    JP
    DE
    PL
    IN
    CN
}
```

If you really wanted to, you could add country descriptions; however, it's not necessary.

On the other hand, take a look at this enum:

```
enum Status{
    PENDING
    OPEN
    STORE
    CLIENT
    DELIVERED
}
```

In such cases, we can only guess what the schema author had in mind. Does STORE mean a shop here or a warehouse? Or does it mean that we need to store something somewhere? That's why GraphQL allows us to document enum fields, so we can explain to users what each status represents:

```
"""
Show the current status of the order.
"""
```

```
enum Status{
    """
    Order is pending payment
    """
    PENDING
    """
    User has a cart with their order opened.
    """
    OPEN
    """
    Order is ready to be sent
    """
    STORE
    """
    Order has been sent to a client
    """
    CLIENT
    """
    Order is delivered
    """
    DELIVERED
}
```

Perhaps the order statuses in this example are not well-constructed (such cases will often occur when creating a REST proxy – a GraphQL schema created only for purposes of proxying under the hood REST system – from another system); however, these minimal pieces of information will greatly facilitate work and reduce misunderstandings during implementation or consumption of the schema.

Unions

Unions are much simpler than interfaces – they allow for returning one of the types belonging to the set of possible types, but they do not guarantee any shared fields. The types within a union are not aware of each other, unlike interfaces where the `implements` keyword defines that relationship.

Take a look at this example:

```
type Query{
    search(query:String): [Result!]
}

type Person{
    firstName: String!
    lastName: String!
}
```

```
type Company{
  name: String!
}


type Group{
  groupName: String!
  category: String!
}
union Result = Person | Company | Group
```

In this example, from the search, we can see that the returned `Result` type can be one of three types: `Person`, `Company`, or `Group`

When we make a search query, it is essential to be able to differentiate the results based on the queried type. This is where the `__typename` field comes into play. The `__typename` field always returns the name of the queried type in the result, so we can get the actual type from unions and interfaces.

Using the `__typename` field provides us with valuable information about the type of data we are dealing with, allowing us to handle the data differently based on its type, and enabling us to implement specific logic and functionality for each type.

Let's look at how to use the `__typename` field in a GraphQL query:

```
{
  search{
    __typename
    ... on Person{
      firstName
      lastName
    }
    ... on Company{
      name
    }
    ... on Group{
      groupName
      category
    }
  }
}
```

This GraphQL query retrieves information from the `search` field. The query includes inline fragments that allow the conditional selection of fields based on the type of the object being queried.

Inline fragments are used to specify different fields and their values depending on the type of the object returned by the search field. In this query, there are three inline fragments defined using the ... on TypeName syntax. The first inline fragment, ... on Person, specifies that if the object returned by search is of the Person type, then the firstName and lastName fields should be included in the response. These fields represent the first name and last name of a person.

By using inline fragments, the query can conditionally select different fields based on the type of object being queried, allowing for more dynamic and flexible data retrieval.

Here is the response structure:

```
{
  "search": [
    {
      "__typename": "Person",
      "firstName": "Jerrod",
      "lastName": "Daniel"
    },
    {
      "__typename": "Company",
      "name": "Aexol"
    },
    {
      "__typename": "Group",
      "groupName": "Cat lovers",
      "category": "Animals"
    }
  ]
}
```

When fetching unions or interfaces from the backend, we need to remember to include the __typename field. Otherwise, we would have to duplicate the work of the backend developer on the frontend and manually differentiate between types.

Interfaces

Interfaces in GraphQL have an interesting feature that requires additional documentation. Specifically, if an interface is returned on a field of a certain type, it means that the field can return any of the types that implement the interface. When I was developing the graphql-zeus open source library, this was the biggest challenge for me in mapping GraphQL types to TypeScript.

In the GraphQL specification, it is the types that implement the interface, not the interface that specifies which types it covers. Therefore, additional documentation for the interface is useful so that users who do not have a GraphQL schema visualizer can immediately see what will be returned from the backend on a specific field that returns an interface.

Let's start by reminding ourselves how interfaces work in GraphQL using a very small schema as an example. This will immediately highlight the need for documenting them:

```
interface Animal{
  name: String
  continent: String
}

type Query{
  animals: [Animal!]!
}

type Reptile implements Animal{
  camouflage: Boolean
  name: String
  continent: String
}

type Fish implements Animal{
  swimmingSpeed: Float
  name: String
  continent: String
}

type Dog implements Animal{
  breed: String
  name: String
  continent: String
}
```

Here, we have the Dog, Reptile, and Fish types, implementing the Animal interface. This is pretty clear, but I will also add some visualization to it:

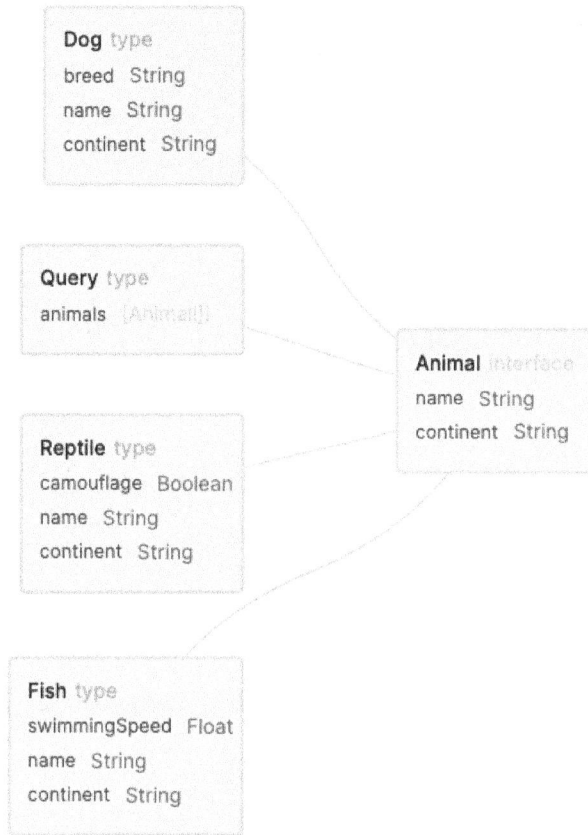

Figure 1.3: Visual schema graph of the Animal interface

Now imagine you can only see Query.animals because our schema is an encyclopedia schema. Without visualization or documentation, you would only see that it returns Animal with name and continent. You could use tools to visualize or query the schema you can get all the types, but adding just small documentation to the Animal interface can also help here:

```
"""
Animal interface responsible for returning animal base data. It
returns Dog, Reptile or Fish.
"""
interface Animal{
  name: String
  continent: String
}
```

After adding documentation, the user can at least see the possible types before using a tool that provides a language server on GraphQL schema.

Additionally, interfaces can be queried in the same way as unions, with the exception that we can query the fields that are common can be queried together. This example shows how to query interfaces:

```
{
  animals{
    __typename
    name
    continent
    ...on Reptile{
      camouflage
    }
    ...on Fish{
      swimmingSpeed
    }
    ...on Dog{
      breed
    }
  }
}
```

As you can see, we just queried three different types using the `interface` field. This allows us to return different type objects in one array from the backend without losing the type information.

Inputs

When a field argument requires a more complex set of input parameters, we use **inputs** to define its structure. By adding inputs to the schema, we can improve its overall readability and organization. This can be seen when comparing the `addBook` and `lendBook` mutations:

```
type Mutation {
  addBook(
    book:CreateBook!
  ): Book!
  updateBook(
    id: ID!
    book: EditBook!
  ): Book!
  deleteBook(id: ID!): Book!
  lendBook(
    id: ID!,
    borrower: String!,
    dueDate: String!
    ): Book!
  returnBook(id: ID!): Book!
```

```
}

input CreateBook{
  title: String!
  author: String!
  genre: String!
  publicationYear: Int!
}

input EditBook{
  title: String
  author: String
  genre: String
  publicationYear: Int
}
```

Here, we moved all the arguments used on the `Mutation` fields to the dedicated `input` nodes. Let's take a closer look.

In the `addBook` mutation, the user is required to provide the full object as an argument. This means that all the necessary information, such as the book's title and author, needs to be included in a single object. While this approach works, it can become cumbersome and less intuitive, especially when dealing with complex objects.

On the other hand, in the `lendBook` mutation, individual parameters are used instead of a single object. This allows the user to provide specific information, such as the book's ID and the borrower's ID, as separate arguments. This approach offers more clarity and simplicity when performing the mutation.

By utilizing inputs in GraphQL, we can enhance the schema's structure and readability. It provides a more organized and intuitive way to pass arguments, making it easier for developers to understand and work with the API. So, even though both approaches achieve the same functionality, using inputs can significantly improve the overall schema design.

Now, we can use those mutations and provide variables of the `CreateBook!` or `EditBook!` types. This adds clarity to the schema and allows us to write more readable queries by using variables.

At this point, we have highlighted the importance of documenting GraphQL schemas. Now, let's move on to GraphQL operations and explore different ways of executing it.

Understanding the GraphQL query language

As I mentioned earlier, introspection is vital, and now I present to you the most important part that benefits from introspection: the **GraphQL query language**. While REST APIs allow for exposing the JSON schema, which provides information about the API structure, I have rarely encountered good use cases for it throughout my professional career. On the other hand, the GraphQL query language is widely used in a proper manner because the language makes it hard to make mistakes.

When I first started using GraphQL, I had this feeling of "*Why didn't people agree on such a standard earlier?*" We didn't need artificial intelligence or quantum computers for it – all we needed was to communicate and reach a consensus – but now GraphQL is commonly used and it is a standard we needed. While it may not be visibly apparent at this moment, I can confidently say that GraphQL is pushing programming technologies forward significantly.

In this section, we will focus on understanding how our schema defined using the SDL is transformed into an AST, which allows communication with the system hidden behind the GraphQL SDL using the GraphQL query language.

To do this, we'll return to our library schema and see what we can achieve by querying the following schema:

```
type Query{
    """
    Get all books that are in the library, including borrowed
      books
    """
    getAllBooks: [Book!]!
    """
    get book by its database id
    """
    getBookById(
        id: ID!
    ): Book
    """
    Get all the books that are physically in the library.
    """
    getAvailableBooks: [Book!]!
}
```

> **Note**
>
> I won't build a backend for now – I will rely on mocks – however, stay tuned, as there will come a time when we will build a fully functioning GraphQL server!

Let's start by trying to get information using `Query.getAllBooks`. We will fetch basic book information, as well as fetch an author and their details in one query:

```
query ListBooks {
  getAllBooks {
    author {
      firstName
```

```
            lastName
            photoUrl
        }
        coverUrl
        genre
        id
        publicationYear
        title
    }
}
```

It is optional to give a query a name; however, naming a query is very useful because it allows us to have multiple operations with different names in one file. The name of the query here is ListBooks.

The query is typically sent as a POST request with the following body:

```
{
    "query": "...",
    "variables": { ... }
}
```

The query field contains the GraphQL query itself, while the variables field holds any variables that need to be passed along with the query.

Here is the result of the query:

```
{
    "getAllBooks": [
        {
            "author": [
                {
                    "firstName": "Ryan",
                    "lastName": "Johnston",
                    "photoUrl": "http"
                },
            ],
            "coverUrl": "Taiwan",
            "genre": "female",
            "id": "Bicycle",
            "publicationYear": 1,
            "title": "Wilkinsonfield"
        },
        {
            "author": [
                {
```

```
        "firstName": "Yasmine",
        "lastName": "Oberbrunner",
        "photoUrl": null
      },
    ],
    "coverUrl": "Dominican Republic",
    "genre": "male",
    "id": "Union",
    "publicationYear": 3,
    "title": "West Jordan"
  }
  ]
}
```

As you can see, the results in GraphQL are in JSON format.

Thanks to the built-in introspection in GraphQL, this powerful feature allows us to dynamically fetch the schema and use it to enhance our development experience. By retrieving information about the available types, fields, and arguments, we can provide autocomplete suggestions and validate our queries against the schema, which greatly reduces the chance of errors and makes our code more robust.

Also, with the help of introspection, we can explore and understand the capabilities of the GraphQL API without relying on external documentation or guessing the structure of the data. This not only saves time but also improves collaboration between frontend and backend developers as they can easily synchronize their work based on the shared schema definition. Instead, we get all the information about our resolvers from the /graphql endpoint.

To explain this a bit further, when writing a query to our backend, we usually don't know the exact type of response we will receive until the query is executed. However, with GraphQL, we can be certain about the type of object that will be returned from the backend because GraphQL enforces the type of that object. If the backend returns an incorrect type, the response will not be received and an error will be displayed instead.

At this point, we have described one of the key goals of GraphQL technology, which is built-in type safety from the beginning. Now, let's move on to exploring the additional capabilities of the GraphQL query language.

Fragments

The first capability we will look at is fragments. Now, what exactly is a fragment? We briefly met this topic earlier, but a **fragment** is an optimization for both the transmitted query and the written query. It allows us to avoid repetition when requesting books in multiple places within a single query, reducing the size of the transmitted query and improving its readability.

In this code block, we will create a fragment for Book:

```
fragment bookDetail on Book{
   author {
      firstName
      lastName
      photoUrl
   }
   coverUrl
   genre
   id
   publicationYear
   title
}

query ListBooks {
   getAllBooks {
      ...bookDetail
   }
   getBookById(id:"h3784rh2873r"){
      ...bookDetail
   }}
```

As you can see, by using the `fragment` keyword, we don't have to repeat the fields we selected for the Book type twice.

But let's remember that in GraphQL, there are also aliases, and it is precisely those that enhance the usefulness of GraphQL fragments as we can reuse the same selectors to get the same response format for different objects fetched individually.

Aliases

By default, GraphQL-based server response fields have the same name as we requested in GraphQL. But what should we do when we want to execute the same field, for example, the child of the root query, multiple times in one query? That's where aliases come to the rescue.

An **alias** is a way to change the name from the original query in the response, which allows us to request the same fields multiple times. Of course, this only makes sense when the field accepts some parameters.

In this query, we want to fetch three specific books with one query:

```
query ListBooks {
   pocahontas:getBookById(id: "4sda38731aa"){
      ...bookDetail
   }
```

```
robinsonCrusoe:getBookById(id: "8u12h3871h3"){
  ...bookDetail
}
treasureIsland:getBookById(id: "8u12aa87fsd"){
  ...bookDetail
}}
```

Now, in the query response, it no longer includes the `getBookById` field but, instead, has specific aliased fields containing results for each book:

```
{
  "pocahontas":{...}
  "robinsonCrusoe":{...}
  "treasureIsland":{...}
}
```

Aliases are very useful because they do some of the work for us from the perspective of a frontend developer. Normally, we would need a `getBooksByIds` resolver that takes an array of IDs, and then we would have to manually sort the response and match each ID to the required field, but here, aliases do that for us. This is very useful when we want to execute multiple identical queries with different parameters and have named responses for each parameter.

Arguments

Arguments in GraphQL allow us to create parameterized functions that can accept data as input. This enables us to customize the behavior of our GraphQL queries and mutations based on specific criteria or requirements. By providing arguments to a field, we can pass in values that will be used by the resolver function on the server side to perform certain operations or fetch specific data. By leveraging arguments in GraphQL, we can build more dynamic and adaptable APIs that cater to a wide range of use cases and scenarios. Of course, in GraphQL, we can execute functions and pass parameters to the backend, just like in REST.

Arguments can be defined as scalars, enums, and inputs. In the following example, we will see how arguments are defined in the GraphQL SDL:

```
type Mutation{
addBook(
  book: CreateBook!
  ): Book!
}
```

The word book here is an argument, meaning that the user needs to pass it within the query or as a variable.

> **Note**
>
> You may be wondering whether a union or an input union can be used as a parameter. Well, in GraphQL, such a thing does not exist, and you are required to define an exact input or scalar.

Variables

Yes, GraphQL also includes variables. They are used to avoid putting large inputs directly into our query and instead send them separately in the `variables` field. Therefore, we have two mechanisms by which we can pass parameters to a function: standard insertion in the query or using variables.

Using our `addBook` mutation, let's see how to use variables in GraphQL. First, let's look at the schema:

```
input CreateBook{
   title: String!
   author: String!
   genre: String!
   publicationYear: Int!
}
type Mutation{
addBook(
   book: CreateBook!
   ): Book!
}
```

Next, we can write the query against that schema and send variables as a separate object:

```
mutation AddBook($Book: CreateBook!) {
   addBook(book: $Book) {
      id
   }
}
```

Variables are then sent as a separate field and injected into the query inside the backend system. This makes the query itself easier to read without inserting the whole object inside the query:

```
{
   "Book": {
      "genre": "Fantasy",
      "title": "Lord of the rings",
      "publicationYear": 1954,
      "isAvailable": true,
      "coverUrl": "https//example.com/lotr.png"
   }
}
```

Here is the JSON object we will be sending in the `variables` field. We defined the `Book` variable and added fields to it.

Later, the query and variables are sent in the JSON format:

```
{
    "query": "GQL QUERY HERE"
    "variables": { ...VARIABLES HERE }
}
```

As you can see, GraphQL variables play a crucial role in sending data and analyzing query performance. By using variables, we can easily group together multiple queries with different inputs, making it simple to identify and evaluate their individual performance. This powerful mechanism enhances the clarity and efficiency of working with GraphQL, ultimately improving the overall data management process.

Directives

Directives in the GraphQL query language tell us how the communication with the server should be conducted and which fields should be included or skipped. They provide instructions on how the query should be executed and allow us to control the behavior of the server-side execution.

Here, we will briefly describe the most important built-in directives:

- `@include`: This directive allows for the conditional inclusion of fields in the response based on a `Boolean` condition. It takes an argument called `if` that determines whether the field should be included or not. Used in query.

- `@skip`: Similar to `@include`, this directive also controls the inclusion of fields in the response. However, it skips the field if the `Boolean` condition is `true`. Used in query.

- `@deprecated`: This directive is used to mark a field as deprecated. It provides information about the deprecation and suggests an alternative field or solution.

- `@defer`: This directive is used to delay the execution of a field until the rest of the query has been resolved. It is useful for optimizing performance by fetching non-critical data later. Used in query.

- `@stream`: This directive is used to stream a field's results as they become available, instead of waiting for the entire field to resolve before sending the response. Used in query.

These are just a few examples of directives in GraphQL queries. Each directive serves a specific purpose and allows for more flexibility and control over the query and its response.

Response format

You would be right in thinking that the correct response format in GraphQL is regular JSON. However, it's important to note that the response format significantly differs from REST.

In a REST API, when we make a request to an endpoint, we receive a predefined payload containing all the data related to that particular endpoint. This means that we often receive more data than we actually need for a given use case. This can result in unnecessary network overhead and increased response times, especially when dealing with large and complex APIs.

In comparison, the response format of GraphQL is structured in a hierarchical manner, mirroring the structure of the query. This means that the response object will have the same shape as the query object, making it easier to traverse and work with the returned data.

In most cases, the response will have a status of 200 and may contain fields such as data and errors. The data field holds the requested data, while the errors field contains any errors encountered during the execution of the query.

Of course, responses with different status codes can also occur, but they usually indicate errors at a different level, such as network or server errors. It's crucial to handle these different response formats appropriately when working with GraphQL APIs.

Workarounds for input unions

GraphQL does not have built-in support for input unions but, instead, we can use several different methods to mimic this functionality to some extent (I won't criticize the lack of this feature here, as I really appreciate GraphQL for its simplicity).

Let's just look at one method. We can create dedicated fields for input unions in cases where we need multiple input types. This helps us avoid runtime errors because each field is prepared to handle the set of arguments defined in the schema. Here is an example:

```
type Query {
  getBook: GetBookQuery
}

type GetBookQuery {
  getBookById(id: ID!): Book
  getBookByTitle(title: String!): Book
}
```

This approach ensures that the input is validated against the defined arguments at compile time, reducing the chances of encountering errors during runtime.

Now, to wrap up this section, we have learned what the GraphQL query language exactly is, including how to create queries and some more advanced applications such as aliases. This knowledge will definitely come in handy in the subsequent parts of the book. In the final section, we will explore how to mock responses even before the backend is ready.

Mocking GraphQL responses

Mocks refers to data that is used when we don't have the system implementation yet. Their purpose is to closely resemble the data that will be returned from a functioning backend.

But how do mocks work? It's quite simple. Each type, except for custom scalars, ultimately resolves to basic scalars such as `String`, `ID`, `Float`, `Int`, and `Boolean`. By knowing the final type and the resolver path, sometimes, even with detailed documentation, we can return very accurate mock values.

Let's take a look at our simplified `Book` type:

```
type Book {
    id: ID!
    title: String!
    author: String!
}

type Query{
    books: [Book!]!
}
```

If the fields are built-in scalars, we are able to provide mocks just after looking at the type. In the following JSON code, we provide a response for `Query.books`:

```
[
    {
        "id": "123abc",
        "title": "Robinson Crusoe",
        "author": "Daniel Defoe"
    },
    {
        "id": "939301i",
        "title": "Treasure Island",
        "author": "Louis Stevenson"
    }
]
```

As you can see, this JSON precisely represents how mocks are done in GraphQL. Later in the book, we will also see what AI prompts can help us generate more accurate mocks, too.

Now, before we end this chapter, let's set up your first GraphiQL instance with your own GraphQL schema and mock backend. Here are the instructions:

1. Open your command line.

2. Create a new folder (called `myfolder`) and enter the folder.

3. Make sure to have Node.js and npm installed.

4. Then, install GraphQL Demeter globally:

```
npm install -g graphql-demeter
```

   This will install the mock engine for GraphQL on your computer.

5. Next, create the `schema.graphql` file with the following content:

```
type Query{ hello: String! }
```

6. Now, run the mock engine with the following command:

```
demeter schema.graphql
```

7. Your mock server should be running on port `4000`. Go to `http://localhost:4000` in the browser and you should see this screen:

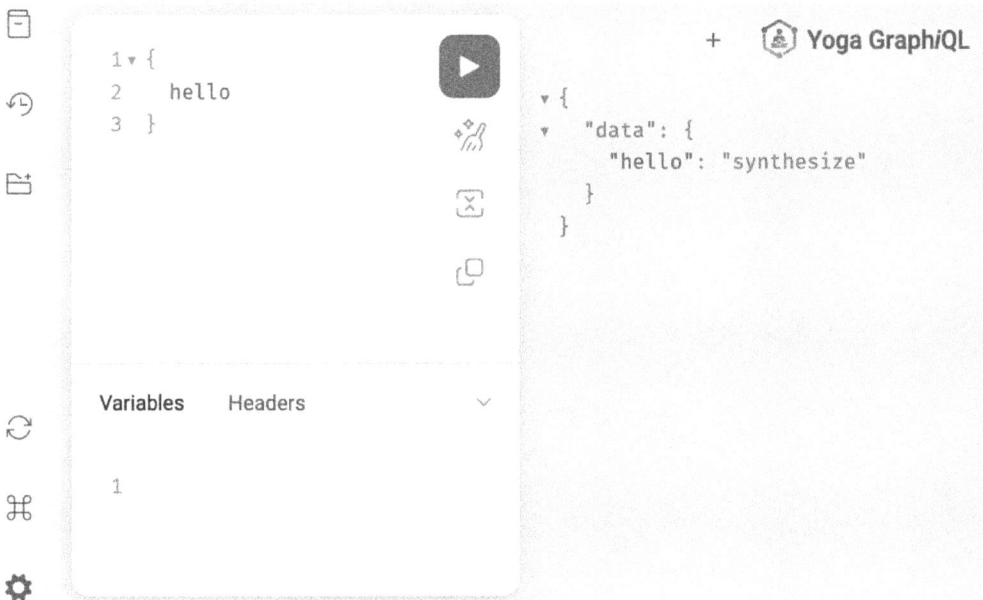

```
1 ▾ {
2     hello
3   }
```

```
▾ {
▾   "data": {
      "hello": "synthesize"
    }
}
```

Yoga GraphiQL

Variables   Headers

```
1
```

Figure 1.4: GraphiQL interface

This is the GraphiQL interface. You can enter queries here to receive responses.

Now, you should modify the schema and mock value configurations to fully play with GraphQL and GraphiQL. The documentation of this small GraphQL mock engine can be found here: `https://github.com/graphql-editor/graphql-demeter`.

Summary

In this chapter, we have reviewed the basics of GraphQL and highlighted its greatest strengths. We have learned how to query the GraphQL backend for its structure and how it is defined in the SDL. These are all solid foundations of GraphQL that will allow us to smoothly progress through this book. If you don't understand something in the upcoming chapters, you can always come back here.

Before we dive into the heavyweight schema definition in GraphQL, I'd like us to take a look at how GraphQL relates to large language models and how these two technologies can work together seamlessly. In the next chapter, you will learn how to make GraphQL schemas AI-friendly.

Applying an AI-Friendly Approach to GraphQL

Today, **Artificial Intelligence** (**AI**) has dominated the world. We use AI to search the internet and find answers to the most pressing questions. We can use it to create virtual assistants for various purposes and generate beautiful images. The whole world is captivated by this technology.

However, what does this have to do with GraphQL, you ask? Well, a lot. GraphQL is a language that was designed to facilitate communication within programming teams, and as it turns out, it also excels at improving human-computer communication.

In this chapter, we will learn how to leverage AI to accelerate our work with GraphQL schema-based systems. After reading this chapter and adding AI tools to your arsenal, you will undoubtedly work much faster with GraphQL.

In this chapter, we will cover the following topics:

- Understanding GraphQL AI friendliness
- Navigating the OpenAI playground
- Generating documentation
- Generating real-life mocks
- Writing GraphQL queries with AI
- Creating whole schemas

Technical requirements

To follow along with this chapter, you need access to the OpenAI playground with the **gpt-4** model, which you can find at `https://platform.openai.com/playground` (make sure to create an account on OpenAI to create and use the AI systems).

Alternatively, you could use other conversational tools such as Llama and Gemini.

You can also find the code for this chapter on GitHub: `https://github.com/PacktPublishing/GraphQL-Best-Practices/tree/main/chapter-02`.

Understanding GraphQL AI Friendliness

GraphQL and REST are two different API architectures that offer different approaches to communication between client and server. In the world of web development, understanding the fundamental differences between the two is crucial. One key aspect that sets these two paradigms apart is the way they handle URL paths.

In REST, URL paths are used to define the resources that the client wants to interact with. For example, if we have an endpoint called `/users`, it typically represents the collection of all users in the system. To access a specific user, we would append their unique identifier to the path, resulting in something such as `/users/123`. This hierarchical structure helps organize the API and provides a clear representation of the data model.

On the other hand, GraphQL takes a different approach. Instead of using a fixed path structure, GraphQL has a single endpoint, often `/graphql`, to which all requests are sent. The client sends a query or mutation along with the necessary parameters, and the server responds with the requested data. This means that the entire API is exposed through a single URL, eliminating the need for multiple endpoints.

Here, the focus shifts from the URL path to the query itself. The query language allows the client to precisely specify the data it needs, reducing the over-fetching or under-fetching of data that is common in REST APIs. By defining a query using fields and nested selections, developers can efficiently retrieve only the required data in a single request. Due to this, GraphQL is easier to consume for AI engines compared to REST.

Let's break down the differences between GraphQL and REST, and how they relate to AI friendliness, a little more:

- **Single entry point**: In REST, each resource has its own unique URL, which means you have to make multiple HTTP requests to fetch all the necessary data. With GraphQL, the client sends a single query to a single entry point, and the server returns only the requested data. In this way, AI has a single point of communication and doesn't need to have knowledge of existing endpoints, which it would only get from the documentation in the case of REST.

- **Flexibility**: REST is based on a fixed set of endpoints provided by the server. With GraphQL, the client has control over the data it receives by defining its own queries. This provides greater flexibility in fetching only the required information. AI can construct a precise query and retrieve only the necessary information.

- **Documentation**: GraphQL has built-in and automatic documentation generated based on the GraphQL schema. This makes it easier to understand the API structure, available fields, and relationships between them. This also saves developers a significant amount of time and effort, as they no longer need to manually write and update documentation themselves.

So, thanks to these differences, GraphQL makes it easier for AI to understand the way the API works.

Next, we will move on to OpenAI, which we can use to create **AI assistants**.

Navigating the OpenAI playground

As mentioned in the *Technical requirements* section, to play with all the AI examples in this chapter, you need an OpenAI account to access their interactive playground.

After creating the account, you can access the playground that we will use in this chapter at `https://platform.openai.com/playground/assistants`. The playground should look like this:

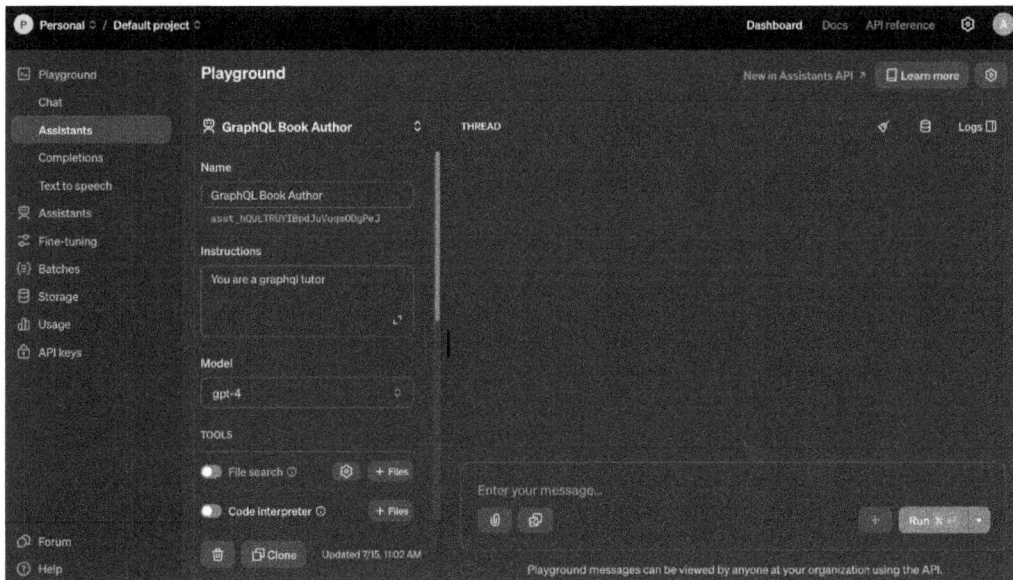

Figure 2.1: The OpenAI playground

From here, you can create an AI assistant (also known as a chatbot) – this is a virtual entity powered by AI that is designed to assist and support individuals with various tasks, providing personalized and intelligent responses based on user input.

To create and use the AI assistant, the main parts of the interface you need to know are as follows:

- **Name**: This is where you enter the name of the assistant.

- **Instructions**: This is where you enter the system message for your assistant, used to specify its behavior.

- **Model**: This is where you choose from the different models. Each model has different functionalities. We will use **gpt-4**, which is a text generator.

- **Enter your message…**: This is where you can talk and give prompts to the assistant.

Now that you understand how the OpenAI interface works, let's create an AI assistant to generate documentation.

Generating documentation with AI

Writing documentation for GraphQL APIs can be a time-consuming task. However, AI can simplify this process by analyzing the schema and automatically generating documentation based on it. By understanding the structure of the API and its various fields, AI algorithms can generate detailed and accurate documentation, reducing the manual effort required from developers. We will base all our AI functions on **Large Language Models (LLMs)**.

An LLM is a powerful AI tool that has the ability to generate text and understand natural language. It is a language model that has been trained on vast amounts of textual data to learn various language patterns and structures. In the context of this book, LLMs will be used to automatically generate GraphQL Schema Definition Language code, mock responses, documentation, and GraphQL Query Language.

Let's use the library example from *Chapter 1* and try to generate documentation for our Book type. Of course, an LLM won't be able to figure out how to do it on its own. That's why we'll provide it with a system message that will define its role.

Within the system message, we need to provide the role, input prompt format, and response format. Based on the provided system message and prompt, the system analyzes the context and tries to understand the user's intention. Then, using the knowledge gathered during training, the model generates a response that is consistent with the input data.

The response from the LLM system is generated in the form of text. It can be a single sentence or even multiple sentences, depending on the complexity of the task. The response is created to be logical and relevant to the information provided in the message and command.

Let's sum up what we said here so we can easily refer back to it:

- **System message:** This is the message that specifies the role of the LLM model. The model then processes the input using this system message for context and generates a response that is coherent and contextually relevant.

- **Prompt:** This is the message sent to the LLM that is used to produce a response.

> **Note**   .
>
> The following LLM-generated definition is a perfect explanation of prompt engineering: "Prompt engineering is the art of crafting user-friendly and efficient prompts for developers to interact with GraphQL APIs, ensuring a seamless and intuitive experience. It involves designing and implementing query structures, variables, and documentation that empower developers to easily explore and retrieve the data they need."

- **Response**: This is the response from the LLM in the format specified in the system message.

It is important to remember that the responses generated by an LLM are based on the data it was trained on. This can lead to certain limitations, especially if the system lacks sufficient experience in a particular area. Therefore, it is important to appropriately customize and verify the responses to ensure they align with the user's expectations.

Great! Now let's dive into generating documentation for individual nodes in a GraphQL schema. To begin, we'll focus on generating documentation for GraphQL types and their corresponding fields.

With the OpenAI playground open, create an AI Assistant called `GraphQL Documentation Tool`. Then, in the **Instructions** box, enter the following system message:

```
You are a GraphQL Documentation Tool. You receive a GraphQL schema and
you return a documented GraphQL schema in the response. For example,
if you receive schema:
```graphql
type TYPE_NAME {
  FIELD_NAME: FIELD_TYPE
}
```

You should respond with:

```graphql
"""
description of type
"""
type TYPE_NAME {
  """
  description of field
  """
  FIELD_NAME: FIELD_TYPE
}
```
```

In this message, we explained how documentation works in GraphQL to the LLM:

- First, we input the message with the role of the system message, saying it will receive a GraphQL Schema
- Second, we specified the format of the message LLM will receive
- Third, we specified what the response structure should look like

Of course, we could write a much longer and more detailed system message, taking into account the entire GraphQL specification. However, for the purpose of this example and for the sake of generating results quickly, this simple system will suffice.

> **Note**
>
> I intentionally used special markers of three tildes (```) and the word `graphql` (so ```` ```graphql ````) because the language in which we communicate with LLMs and write the system message is **Markdown**. Most AI systems use Markdown as it is easy to embed other languages in Markdown responses. Without Markdown, the LLM would need to decipher what is GraphQL and what is plain text.

Now that our system message is set and the GraphQL Documentation assistant knows what it is doing, in the **Enter your message…** section, let's enter the following prompt to send it to LLM:

```graphql
type Book {
  id: ID!
  title: String!
  author: String!
  genre: String!
  publicationYear: Int!
  isAvailable: Boolean!
  borrower: String
  dueDate: String
}
```

In this example, we wrapped the GraphQL language with Markdown.

Here is the result:

```graphql
"""
A book in the library.
"""
type Book {
```

```
"""
The unique identifier for the book.
"""
id: ID!

"""
The title of the book.
"""
title: String!

"""
The author of the book.
"""
author: String!

"""
The genre of the book.
"""
genre: String!

"""
The publication year of the book.
"""
publicationYear: Int!

"""
Indicates if the book is currently available.
"""
isAvailable: Boolean!

"""
The name of the person who borrowed the book (if any).
"""
borrower: String

"""
The due date for returning the book (if applicable).
"""
dueDate: String
}
```

Voila – this is the response our GraphQL-trained LLM provides.

You can also see what the response looks like in the OpenAI playground here:

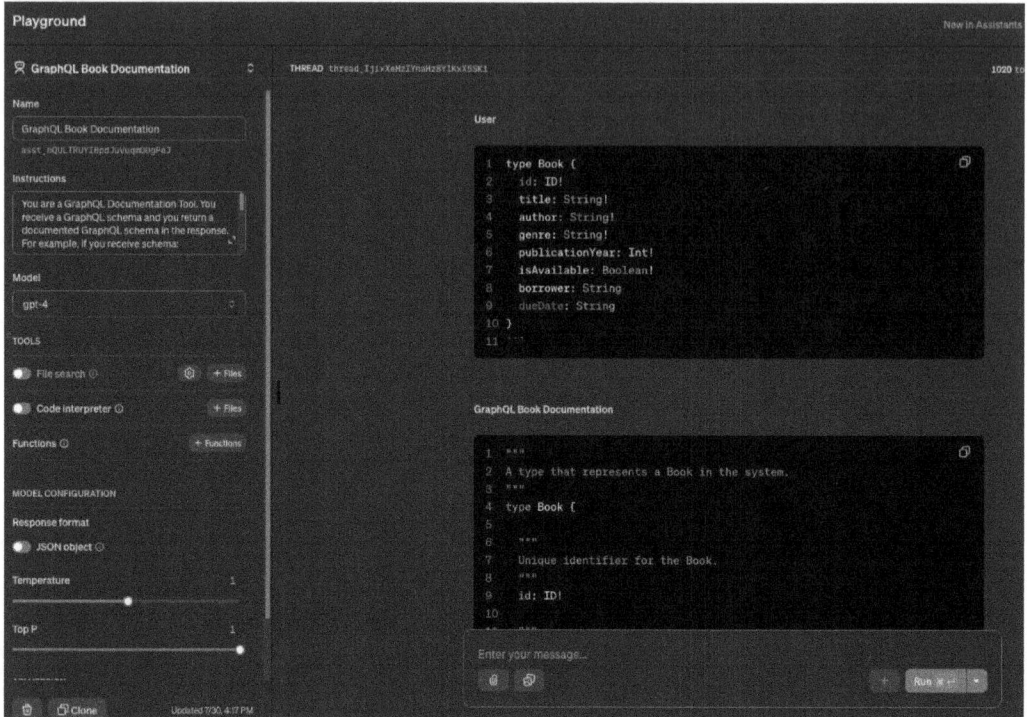

Figure 2.2: The response from the LLM

We no longer need to document anything, as everything will be done for us. Of course, there are cases where the LLM may not describe a type exactly as we want, but in such situations, we should refine the system message and try again, or even use fine-tuning on the AI model itself. So, always make sure to double-check the results yourself.

Now that we have a well-documented schema, it can be better handled by an AI system that generates mocks (based on my experience, I can tell you that starting a project with a well-prepared mock backend significantly speeds up the process).

Generating real-life mocks

By analyzing schemas and understanding the structure of APIs, AI algorithms can generate realistic mock data that can be used for testing or development purposes. This saves time and effort for developers, as they no longer need to manually create mock data.

So, let's create a system message that explains how the LLM should respond and what data it should include when generating mock data. In the OpenAI playground, create another AI assistant called **Mock Object Generator**, and add this system message to the **Instructions** box:

```
You are a Mock Object Generator. You receive a GraphQL schema and you
generate an array of mock objects in JSON format.
```

This message is much simpler than the prompt we used in the previous section – this time, it doesn't need more data as LLMs usually know JSON format.

Now let's use the code previously generated at the end of the *Generating Documentation with AI* section as the prompt. This is a great example of how we can combine AI tools in pipelines. By using the code with documentation, we can create more accurate mocks so the LLM doesn't have to process the information and figure out what each field means. This integration of AI tools can greatly enhance our development process and improve the quality of our applications.

So, after giving the code as a prompt, we will receive the following response:

```json
[
    {
        "id": "1",
        "title": "To Kill a Mockingbird",
        "author": "Harper Lee",
        "genre": "Fiction",
        "publicationYear": 1960,
        "isAvailable": false,
        "borrower": "John Doe",
        "dueDate": "2022-05-15"
    },
    {
        "id": "2",
        "title": "1984",
        "author": "George Orwell",
        "genre": "Fiction",
        "publicationYear": 1949,
        "isAvailable": true
    },
    {
        "id": "3",
        "title": "Harry Potter and the Philosopher's Stone",
        "author": "J.K. Rowling",
        "genre": "Fantasy",
        "publicationYear": 1997,
        "isAvailable": true
```

```
      }
   ]
   ```
```

Though the actual response contained two more objects, which I removed for brevity, as you can see, the response has got the information right, even including `publicationYear` and `genre`.

Next, we will learn how we can leverage LLM to write GraphQL queries.

## Writing GraphQL queries with AI

Indeed, writing GraphQL queries is not difficult for those familiar with the GraphQL language, and tools such as GraphiQL make it even easier. However, with a deep understanding of the schema and its structure, AI should be able to handle this task as well.

Instead of manually writing a query to receive books by Robert Ludlum, we could simply say, "Hey, give me a list of all books by Robert Ludlum available in the library." Then, under the hood, AI can generate the appropriate query, and GraphQL can return the data accordingly. This demonstrates how AI can assist developers in composing complex queries by understanding natural language instructions.

Here, let's create a message system that includes LLM variables. These variables are written in uppercase letters, and in our case, they are SCHEMA_HERE and PROMPT_HERE – this makes it easier to understand the prompt structure for our LLM system:

```
You are a query generator. You receive schema in GraphQL format and a
prompt. Message format is the following:
Schema:
```graphql
SCHEMA_HERE
```

Prompt:
PROMPT_HERE
You respond with GQL query which is valid according to the schema and
prompt.
```

As you can see, the system first receives the entire schema and a natural language query, and then it composes a GraphQL query from it.

Can this be successful? Let's check by providing the following prompt (note that I am showing only the important part of the schema here):

```
Schema:
```graphql
type Author{
  firstName: String!
```

```
    lastName: String!
    books: [Book!]!
    photoUrl: String
}

type Book {
    id: ID!
    title: String!
    author: [Author!]!
    genre: String!
    publicationYear: Int!
    isAvailable: Boolean!
    borrower: String
    dueDate: String
}

type Query {
    getAllBooks: [Book!]!
    getBookById(id: ID!): Book
    getAvailableBooks: [Book!]!
}

schema {
    query: Query
    mutation: Mutation
}
```

Prompt:
List all books

Prompts don't have to be that long. To reduce the length, we could create a system where the schema is injected underneath the prompt, and only provide the natural language part. We could also embed the schema in the system message, which will give us the same result.

Anyway, here's the response from the system:

```graphql
{
getAllBooks {
    title
    author {
    firstName
    lastName
```

```
}
genre
publicationYear
}
}
```

If we then executed this query, it would give us a list of books.

> **Note**
>
> You may notice that the response is not properly formatted. We would need to improve our prompt to add code formatting to it. Having a consistent prompt format can help improve the accuracy and speed of the AI's comprehension of the user's intent. By defining a clear structure for the prompts, we can ensure that the AI understands the desired actions and generates the appropriate GraphQL queries. To format the response, we can use GraphQL linting libraries and write custom code to run the AI assistant and format the code.

So, by integrating AI algorithms with GraphQL, we can create intelligent automation systems that can automatically perform complex operations based on the data obtained from GraphQL queries. However, now that we know how to communicate with the schema using AI, what if we could generate our ideas as schemas?

Creating whole schemas with AI

What can be surprising is that GraphQL schemas work immediately in GPT systems. They already have a basic knowledge of GraphQL and REST, so they are able to generate definitions for different programming languages.

Let's create a small system to build a GraphQL Schema. Here is the system message that we will use:

```
You are a GraphQL Schema Builder.
```

That's all it takes for the LLM interface to understand what we mean. Of course, we can change the model or use fine-tune it, but for now, this will suffice.

Now let's focus on the prompt. Imagine that we are visionary clients creating a new weather application. This service should provide essential weather data. In this case, the prompt may look like this:

```
A schema for a weather service including city, date, temperature,
wind, rain, and humidity data.
```

These are the basic pieces of information provided by the client. To understand whether the data inside our prompt is adequate, imagine that you are receiving the prompt rather than the LLM. If you understand what kind of application is to be built, the LLM should understand it too.

Now, based on this information, the system builds the schema. Here is the response:

```
type Weather {
  city: String!
  date: String!
  temperature: Float!
  wind: Wind!
  rain: Rain!
  humidity: Float!
}

type Wind {
  speed: Float!
  direction: String!
}

type Rain {
  volume: Float!
  chance: Float!
}

type Query {
  getWeatherByCityAndDate(
    city: String!,
    date: String!
    ): Weather
}
```

This schema is certainly sufficient to start a proof of concept for a new weather application. It includes the basic data requested by the client and is ready for implementation. I recommend experimenting with this model, seeing how you could improve it (for instance, teaching it your way of creating schemas or letting it write resolvers in the programming language of your choice), and connecting it with a basic backend generation system.

Summary

While we may not have built anything flashy such as an AI video generator or a face changer in this chapter, through different examples, we saw how AI can enhance our experience with GraphQL systems, providing capabilities such as mock generation, documentation generation, and AI automation using GraphQL APIs.

Beyond these, the possibilities are endless – you could create a model that provides suggestions to optimize a schema, or a model that detects potential threats in the schema, such as unauthorized access to data through improperly linked graph nodes. This collaboration between AI and GraphQL can streamline development processes and improve overall efficiency.

In the next chapter, we will learn about different schema design patterns.

Part 2 - Schema-First Design Patterns

In this part, we will learn more about GraphQL itself. We will design schemas that will work from the schema-consumer point of view and create GraphQL schemas using access-based patterns.

This part contains the following chapters:

3

Crafting Effective GraphQL Schemas

GraphQL provides its architects with the freedom to create their own schemas, and over the years, I have come across various methods of doing so. However, what I'm about to write here is very important – *errors made at the beginning of schema design will haunt an entire team throughout a project.*

Problems that can arise from a poorly designed GraphQL schema include unauthorized access to data, over-fetching, chaos and messiness, forced execution of a large number of queries instead of one, and unclear relationships between objects. Although GraphQL makes it easy to make changes, they will later require migrations of individual system components. Therefore, proper planning and dedicating time to editing the schema before starting implementation is essential.

In this chapter, we will focus on making sure the schemas we create are easy to use and understand. As we already know, GraphQL – when used as a source of truth – affects the whole team working on a project based on that schema, so you need to take time creating a schema that is compact but well described. That's why we'll explore the rich GraphQL specification of its Schema Definition Language and the rules that you need to stick to to craft the perfect schema.

In this chapter, we will cover the following topics:

- Understanding the basics of schema design
- Harnessing the power of interfaces
- Adding well-documented custom scalars
- Learning about directives

Technical requirements

To complete this chapter, you will need access to a GraphQL IDE such as VS Code, Sublime Text, Notepad++, or GraphQL Editor.

You can access the code in this chapter on the book's GitHub repository: `https://github.com/PacktPublishing/GraphQL-Best-Practices/tree/main/chapter-03`.

Understanding the basics of schema design

In the world of GraphQL, the **Schema Definition Language** (**SDL**) is the foundation on which everything else is built. It is a powerful tool that allows us to describe the structure and behavior of our API clearly and concisely. By using SDL, we can define the types, fields, and relationships that make up our GraphQL schema.

Why is SDL so important? Imagine a scenario where you are building a GraphQL API without any predefined structure. It would quickly become chaotic and difficult to manage. SDL provides us with a standardized way to define our API, making it easier to understand and maintain.

However, it's important to note that SDL comes with certain rules and conventions that we must adhere to. These rules ensure that our schema is valid and can be understood by both the server and the client. Let's take a look at them now.

Standardizing the naming of types and fields

Consistency is a crucial aspect of software development, and standardized naming conventions play a vital role in achieving this within a GraphQL schema.

By using consistent naming conventions, we can easily understand the structure of a schema and its fields. When these conventions are predictable and follow a set pattern, it becomes more intuitive to navigate and comprehend the schema, creating a unified and cohesive appearance, and reducing confusion and potential errors.

This is particularly important in a collaborative environment, where multiple people may be working on the same GraphQL schema. This is because inconsistent naming conventions can introduce unnecessary cognitive load for developers. When field names vary in style or structure, developers may need to spend additional time deciphering each field's purpose and functionality. This can slow down development and increase the chances of introducing errors.

Standardized naming conventions eliminate this cognitive overhead and allow developers to focus on the actual logic and functionality of a schema. It also means that all users are on the same page and have a shared understanding of the schema's structure, facilitating seamless collaboration and reducing the learning curve for new team members joining a project.

Now, let's continue examining our library example from *Chapter 1* and look at a good and bad example of naming conventions in GraphQL. We'll start with a bad example, showing a type called Book:

```
type Book implements Thing{
    name: String
    is_available: Boolean
    createdAt: String
    id: String
}

interface Thing{
    id: String
}
```

Let's break down this example:

- Both kebab case and camel case are used, inconsistently mixing syntax conventions.
- id is of type String instead of type ID – this is because id should be a custom scalar type, as it needs to be encoded/decoded to get its final structure.
- The interface is named Thing, which is not a clear name. It should be easily understood that the type implementing this interface has its representation in the database, so we should change its name to fit this.

All of these points lead to an unreadable schema without any additional information from the author.

Let's try to fix this:

```
type Book implements DatabaseObject {
    id: ID
    name: String
    isAvailable: Boolean
    createdAt: String
}

interface DatabaseObject{
    id: ID
}
```

As you can see, we have addressed each of the issues. Rather than Thing, which doesn't convey any information to the schema consumer, we named the interface DatabaseObject because that's exactly what this interface represents. We have also applied the camel case convention throughout the schema and corrected the type to ID.

Overall, consistency achieved through standardized naming conventions in GraphQL schemas improves readability, understanding, collaboration, and maintenance. It helps developers work efficiently, reduces confusion, and ensures a uniform coding style across an application.

Documenting types, fields, and arguments

The documentation of GraphQL code includes descriptions of types, fields, arguments, and return values. This allows developers to quickly understand how to correctly communicate with a schema and retrieve data from it.

In larger projects, where multiple people work on GraphQL code, documentation serves as a central knowledge hub for an entire team – it enables new team members to quickly familiarize themselves with the functionality and structure of the API, as well as ensuring consistency and unified understanding.

The following example represents the Book type with documentation added:

```
"""
Book represents the base Book object in our system.
"""
type Book implements DatabaseObject {
  """
  The name of the book.
  """
  name: String
  """
  Indicates whether the book is available or not.
  """
  isAvailable: Boolean
  """
  The creation date of the book.
  """
  createdAt: String
  """
  The unique identifier of the book.
  """
  id: ID
}
"""
Implement on every type that has its equivalent in database
"""
interface DatabaseObject{
  """
  The unique identifier of the database object.
  """
```

```
    id: ID
}
```

As you can see, by adding descriptions to the types and fields, we provide extra clarity to the `Book` type and `DatabaseObject` interface.

On top of the previously mentioned benefits, well-documented GraphQL code enables the verification of implementation correctness. Developers can compare documentation with code to ensure that all required fields, arguments, and types are aligned. Documentation can also include query and response examples that aid in testing and verifying whether the API functions as expected.

Designing consumer-friendly schemas

Creating a consumer-friendly GraphQL schema for recipients such as frontend developers makes work much more efficient. This is because it does the following:

- Minimizes the number of queries and returned data, resulting in better performance and faster application operation.

- Facilitates integration with other systems or tools. With clearly defined types and fields, other developers making different tools can connect and integrate with our schema faster.

- Provides all the possible variants of queries and mutations. This means consumers don't have to use the same query with different parameters to receive different backend functionalities and get hundreds of different results; instead, we can provide different queries for different results.

- Allows for the appropriate and understandable naming of fields, types, and arguments, meaning easier API usage and consistency throughout a project.

Now, let's imagine that we could handle object modification in the backend service with a simple `update` function that would do all possible things:

- To update book details, we would just update the fields using `BookInput`.

- To borrow a book, we would change the `isAvailable` field, and in the backend code, we would also need to create a reservation for the book if that field changes.

- To return a book, we would need to change the `isAvailable` field. Then, in the backend code, we would need to check whether the book was borrowed and remove the reservation.

- To change the count of `book` inside `library`, we would need to change the `count` field using `BookInput`.

This is what that function might look like:

```
type Mutation {
  updateBook:(id:ID!, book: BookInput!): Boolean
}
```

Having one function – `updateBook` – is not sufficient or consumer-friendly, as it doesn't convey all the possible actions that can be performed with a book.

Let's see how we can improve this:

```
type Mutation {
  updateBook:(id:ID!, book: BookInput!): Boolean
  borrowBook:(id:ID!): Boolean
  returnBook:(id:ID!): Boolean}
}
```

Now, it is clearly visible to the consumer of the schema what can be done with this `Book` object. Instead of having to guess or ask the schema architect, they can see it has four operations instead of one (we'll still leave `updateBook` as a fallback). Now, the following happens:

- Every time a book is borrowed, the consumer calls the `borrowBook` mutation
- Every time a book is returned to the library, the consumer calls the `returnBook` mutation

Designing a user-friendly GraphQL schema makes it easier to manage changes and expand the API in the future. With clearly defined types and fields, new features can be added, and existing types and fields can be extended without requiring revolutionary structural changes. This accelerates application development and simplifies scalability.

In summary, a user-friendly GraphQL schema brings benefits to both frontend developers and other schema recipients. It facilitates work, simplifies integration, and contributes to application scalability.

Designing a schema by defining relationships before adding other fields

Including relationships at an early stage of the schema design process helps prevent potential issues and inconsistencies. By mapping relationships between different types, we can identify any potential conflicts or challenges that may arise as we add new fields to our schema. This proactive approach saves us time and effort in the long run, as we can address and resolve these issues early on, minimizing the need for extensive schema modifications later.

Another benefit of prioritizing relationships is improved performance. By establishing relationships between data types, we can optimize queries and reduce the number of database hits required to fetch related data. This results in shorter response times and overall system performance improvement. Without properly defined relationships, queries can involve multiple trips to the database, leading to unnecessary delays and decreased efficiency.

Furthermore, adding relationships before other fields facilitates code maintenance and scalability. When we establish clear relationships between types, our schema becomes more modular and extensible. This allows us to easily add new fields or types in the future without causing significant disruptions and breaking existing functionality. It also enables better collaboration within our team of developers, as the schema structure becomes more intuitive and easier to understand.

Only by drawing the relationships can we see how the schema will look and consider whether the relationships have been designed correctly. In the following diagram, we can see the relationships between objects in the library schema:

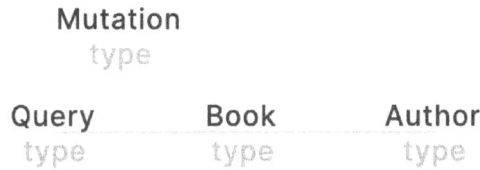

Mutation
type

Query   **Book**   **Author**
type   type   type

Figure 3.1: A relationship diagram of a library schema

The diagram immediately indicates the need for an additional type called BookCopy to store the state of all copies. So, let's update it:

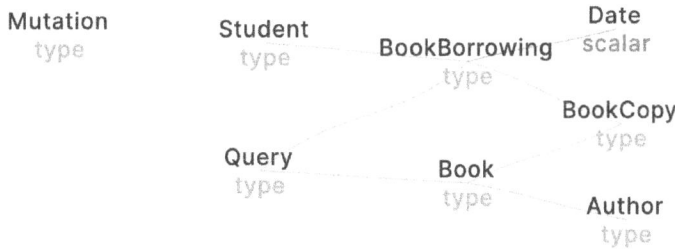

Mutation   **Student**   **Date**
type   type   **BookBorrowing**   scalar
                                                 type

                                                                  BookCopy
                                                                       type
                    Query   **Book**
                       type   type
                                                          Author
                                                             type

Figure 3.2: A fixed library schema diagram

As you can see, we can infer that each book has one or more authors, since it is related to the Author entity.

Moreover, the connection between Book and BookCopy suggests that Book is an abstract object, while BookCopy represents a physical item that is associated with the BookBorrowing object.

It is also important to include the library client in the relational diagram. From the relationship between them, we can see that the Student type is connected with BookBorrowing, so there are some users who can borrow BookCopy by creating BookBorrowing.

> **Note**
>
> This chapter is not about AI, although you can see from the diagram that what is understandable for humans should also be easily understandable for machines (i.e., large language models). In REST, you would usually have the ID of an object returned and have to make another call to the API to get the exact object. However, in GraphQL, that's clear at first glance.

Implementing enums

Using enums allows us to define a specific set of values that a field can take. By using descriptive and meaningful enum values, we can convey the available options in a concise and understandable manner – this makes the schema more self-explanatory and reduces the need for additional documentation or explanations.

Specific enum values also restrict the possible options to a predefined list, reducing the likelihood of errors or misunderstandings when interacting with the API. If a client provides an invalid value for an enum field, the GraphQL response can include a helpful error message, indicating the allowed options. This enables clients to quickly identify and resolve issues, improving the overall developer experience.

In the following example, we can see how an enum is useful for filtering book copies in the library:

```
type Book{
  authors: [Author!]
  copies: [BookCopy!]
}

type Author{
  books: [Book!]!
}

type BookCopy{
  printVersion: String!
  book: Book!
}

type BookBorrowing{
  book: BookCopy!
  borrowed: Date!
  plannedReturn: Date!
}

scalar Date

enum BookCopyStatus{
  """
```

```
    Book is in the library and it is ready to rent
    """
    LIBRARY
    """
    Book is rented.
    """
    BORROWED
    """
    The book is borrowed but not returned on time.
    """
    DUE
}

schema{
  query: Query
}

type Query{
  """
  check for book copies
  """
  bookCopies(
    bookId: ID!
    status: BookCopyStatus!
  ): [BookCopy!]
}
```

In this example, we have added the `BookCopy` type to the schema, which stores a specific copy of a book, as we assume that there can be multiple copies of the same book in the library. We also have the `BookBorrowing` object, which handles each borrowed book. To easily filter books based on their availability, we have the `bookCopies` query, which returns copies of a specific book based on their `BookCopyStatus`. This enum describes all possible states of a book copy.

In summary, it is essential for us to make use of enums whenever possible, especially when the available options are limited. Using enums enhances the understanding of the schema and provides clearer error messages to the recipients of our schema. It improves the readability, maintainability, and usability of our GraphQL API.

Utilizing the @deprecated directive

When adding new fields to a GraphQL schema, it is valuable to use the `@deprecated` directive instead of removing old fields. By deprecating a field, we communicate to clients that it is no longer recommended for use and will be removed in future versions of the API. However, the field remains available for a certain period, giving users time to update their code accordingly.

Using the @deprecated directive helps to manage the migration and evolution of the API. It allows clients to be aware of the deprecation and plan for its eventual removal, while still being able to use the field in the meantime. This approach ensures a smooth transition for users and avoids breaking their existing implementations.

Additionally, the @deprecated directive can provide a reason for deprecation and suggest alternative fields or approaches to achieve the desired functionality. This helps users understand the rationale behind the deprecation and guides them toward the recommended changes.

In the following example, you can see how to use the directive to improve the readability of the schema after changing the name of a field:

```
type Book{
  authors: [Author!]
  @deprecated(
    reason: "Moved to Authors relation fields instead of
    String; this field will be removed"
  )
  authorsList: String
}
```

As you can see, instead of providing authors as a string in the authorsList field, we provide them as a relation to the Author object. We also inform the schema consumer that we are about to remove the authorsList field in the future.

By using the @deprecated directive, we prioritize maintaining backward compatibility and give users the opportunity to adapt to changes at their own pace. It demonstrates a considerate approach toward the users of our API and promotes a smooth and seamless migration process.

Using pagination

As a developer team, we should consider adding pagination only where necessary, critically evaluating whether the standard pagination model can be replaced with another approach. It is important for us to carefully assess the need for pagination in our projects and determine whether it truly enhances the user experience.

The most important features of pagination are as follows:

- Reducing data transfer between the client and server
- Speeding up our project by faster server responses
- Enforcing a good communication style
- Reducing information overload on frontend

However, while pagination can help organize and break down large amounts of content into more manageable chunks, it can also negatively impact user flow and engagement. For instance, end users may become frustrated with having to constantly click through multiple pages to find the information they need. We should be mindful of these drawbacks when considering pagination.

In the following example, I will show you how to implement standard pagination inside GraphQL schema, allowing the schema consumer to limit entries and fetch previous and next entries.

First, we will need a `PageInfo` type to include pagination data inside the response:

```
"""PageInfo contains information about connection page"""
type PageInfo {
  """last element in connection"""
  last: String

  """limit the response size while querying"""
  limit: Int

  """if next field value is false, then client received all
  available data""""
  next: Boolean
}
```

Our strategy is that every response should contain an array of objects, plus the data from pagination.

Having done that, we can create a query resolver that implements this pagination. In this example, we will add `PageInfo` to our books query:

```
type BookConnection{
  books: [Book!]!
  pageInfo: PageInfo!
}

input PaginationFilter{
  limit: Int
  """
  id of the last project returned by previous call
  """
  last: String
}

type Query {
  getAllBooks(pagination: PaginationFilter):
    BookConnection!
}
```

Not only was the shape of the response changed due to the added `PageInfo` type, but we also created a `PaginationFilter` input to allow the schema consumer to use this pagination.

In this example, basic pagination worked well. However, we need to explore alternative approaches that could potentially provide a more seamless and efficient user experience. One such approach is filtering entries by date – the client decides what data from which period is needed, and on the backend, we can limit the maximum period the client can fetch from to prevent sending too much data in one call. Sometimes, that period will be one month, and sometimes, it will be one hour – it depends on how much data is stored. Here is an example with `DateFilter`:

```
scalar Date
input DateFilter{
  start: Date!
  end: Date
}


type Query {
  getAllBorrowings(dateFilter: DateFilter):
    [BookBorrowing!]!
}
```

As you can see, we can now filter books borrowed between certain dates. At first glance, this may seem like filtering, but in reality, we are dealing with a form of pagination, as we can split our objects by the date of creation into chunks. Eventually, we will need those dates if we want to filter those borrowings.

As you continue to develop and refine your GraphQL APIs, keep in mind the specific needs and requirements of your application. Experiment with different pagination strategies, monitor performance, and iterate based on real-world usage patterns. With GraphQL's powerful pagination features at your disposal, you can confidently tackle any data retrieval challenge and deliver exceptional user experiences.

However, delivering those experiences would not be possible without GraphQL's most exciting type – interfaces.

Harnessing the power of interfaces

In the GraphQL language, one of the most interesting features is interfaces. **Interfaces** in GraphQL are designed to help in two main situations:

- They serve as a blueprint for a set of fields that multiple types can implement. By defining an interface, we can ensure that certain fields and their types are present across different object types.

- If you use an interface as a field type, it means that it can return all the types that implement that interface. Moreover, when a type implements an interface, it means that it provides the fields specified by that interface. This allows us to query for objects of different types using a single field.

Let's look at an example. Assuming our library has books and board games available to be borrowed, we will create an interface that brings them together into one collection. Here, you can see how you can return objects of different types in one query:

```
type Book implements LibraryItem{
  authors: [Author!]
  id: ID!
  title: String!
}

type Author{
  books: [Book!]!
  firstName: String!
  lastName: String!
}

  scalar Date

type Query{
  """
  give a list of all library items
  """
  libraryItems: [LibraryItem!]
}

interface LibraryItem{
  id: ID!
  title: String!
}

type BoardGame implements LibraryItem{
  id: ID!
  title: String!
  publisher: Publisher!
}

type Publisher{
  name: String!
}
```

In this example, `Query.libraryItems` returns all the items that are in the library, including items that are of type `Book` and `BoardGame` objects.

Then, we can query this kind of schema as follows:

```
query Items {
  items {
    __typename
    title
    id
    ... on BoardGame {
      publisher {
        name
      }
    }
    ... on Book {
      authors {
        name
      }
    }
  }
}
```

Here, we request items from the library, along with their id, title, and __typename fields. We can query these fields because they are common to both types, and GraphQL knows this because they implement the same interface. Additionally, for the Book type, we request authors, and for the BoardGame type, we request the game's publisher.

But what happens when we want to make our schema more modular? Well, interfaces can also implement interfaces. However, this is not similar to how interfaces work in programming languages, and there are a few important rules to keep in mind:

- First, the type that implements an interface must have the same fields as the interface. Simply using the interface keyword is not enough – the type must also include the same fields as the interface.

- Second, if interface A implements interfaces B and C, then all types that implement interface A must also implement interfaces B and C.

- Third, an interface that implements another interface must copy its fields, just like a type would.

In the next example, we will add a new interface that implements another interface, better specifying the id function across the whole GraphQL schema:

```
interface Node {
  id: ID!
}

interface LibraryItem implements Node {
```

```
    id: ID!
    title: String!
}

type Book implements LibraryItem & Node {
    authors: [Author!]
    id: ID!
    title: String!
}

type BoardGame implements LibraryItem & Node {
    id: ID!
    title: String!
    publisher: Publisher!
}
```

By implementing the Node interface, we have made sure that the function of the interface is not limited to LibraryItem. This decision was made to prevent the id field from being duplicated across different interface objects. Therefore, it is now necessary for every type that implements the Node interface to include the id field with the ID type, ensuring the validity of the schema.

So, implementing interfaces not only allows us to have polymorphic queries but also helps in creating more reusable and modular schemas. We can define interfaces for common sets of fields and have multiple types implement them, making our schema more flexible and maintainable. Additionally, interfaces can be extended by other interfaces, allowing us to create more complex type hierarchies.

In the next section, we will learn about custom scalars, which can provide interfaces for formatted strings, numbers, objects, and even potentially unknown types.

Adding well-documented custom scalars

In GraphQL, custom scalars do not necessarily mean a different underlying type. A custom scalar can still be based on types such as Int or String. So, why is it worth using a custom scalar? It's because custom scalars allow us to avoid repetition and have well-documented fields without duplicating documentation.

When we use a custom scalar, we can define a specific name and behavior for a scalar type that suits our needs. For example, we might have a custom scalar called Date that represents dates in a specific format. Although the underlying type might be String or even Int, using a custom scalar allows us to have a clear and consistent representation of dates throughout our schema.

By using a custom scalar, we can also provide a centralized and well-documented definition for that specific type. This means we don't have to repeat the same description or documentation for every field that uses the custom scalar. Instead, we can define it once and reuse it across multiple fields, improving the maintainability and readability of our schema.

In the following example, we will add custom scalars to our `Book` type:

```
"""
Numeric commercial book identifier that is intended to be unique.
13-digit number
"""
scalar ISBN

"""
Url in format:
https://example.com
"""
scalar URL

type Book {
    id: ID!
    title: String!
    isbn: ISBN!
    website: URL
}
```

Although the `isbn` field type could be written as `Int` or `String`, by creating custom scalars, `URL` and `ISBN`, dedicated to this purpose, we dramatically increase the readability of the schema and understanding of the data format, avoiding chaos.

Additionally, custom scalars can provide extra functionality or validation specific to our application's needs. We can add custom parsing and serialization logic, enforce constraints, or implement specific behavior for our custom scalar. This allows us to have more control and flexibility in handling specific data types within our schema.

Let's take a look at another example:

```
import { GraphQLScalarType } from 'graphql';

new GraphQLScalarType({
  name: 'URL',
  serialize(value) {
    return new URL((
      value as string).toString()).toString();
  },
  parseValue(value) {
    return value === null ? value : new URL(
      value as string);
  },
})
```

In most JS frameworks, you will need to use the GraphQL library's built-in `GraphQLScalarType`. Here, we created a custom scalar function for our URL custom scalar. Using a built-in URL class, it checks if the provided URL is valid. This can be useful, as it adds a validation layer to our GraphQL.

Finally, there is another topic regarding custom scalars that we need to mention. Sometimes, our type is not dependent on the backend system where the schema is defined – in other words, the backend does not have information about the type of object being sent to it, and the type of this object is dynamically dependent on the consumer of the schema.

In such cases, the only option is to use a custom scalar. Although we lose type information at the GraphQL schema level, a custom scalar is the only way to define a dynamic and polymorphic type that is dependent on the consumer. We can see this here:

```
"""
Additional data with different fields for every book, stored in the
String format. Send JSON as string, and parse JSON when received.
"""
scalar MetaDataJSON

type Book {
   id: ID!
   title: String!
   meta: MetaDataJSON
}
```

By having the `MetaDataJSON` scalar inside this schema, we know that we have to stringify it before sending it and parse it when it is returned from the backend.

In summary, using custom scalars in GraphQL allows us to avoid repetition, have well-documented fields without duplicating documentation, and provide additional functionality or validation specific to our application's needs. They provide a way to define and reuse specific types in a consistent and controlled manner, improving the clarity and maintainability of our schema.

In the next section, we will explore the use of directives. While you should be completely fine without them to model the schema for consumers, they may provide some useful functionalities for you as a schema architect.

Learning about directives

While it's not mandatory to use GraphQL directives, they can be incredibly useful in many cases. Directives provide a way to add additional instructions or metadata to our GraphQL schema and queries, allowing us to modify the behavior or control the execution of certain fields or operations.

Handling authentication

One common use case for directives is to handle authentication and authorization. For example, by implementing and applying a directive such as @auth or @hasRole to specific fields or types, we can easily enforce access control rules. This ensures that only authenticated users or users with certain roles can access sensitive data or perform certain operations.

In the following example, you will see how to use directives inside the schema to prevent unauthorized access to the resolver:

```
type User {
  id: ID!
  name: String!
  email: String!
  role: Role!
}

enum Role {
  ADMIN
  USER
}

type Query {
  users: [User!]! @auth
}
directive @auth on FIELD_DEFINITION
```

In this example, we have a User type that specifies the User object in our database or authorization/authentication system. Here, the role field from the User type represents the role of the user and is of the Role enum type, which includes values such as ADMIN and USER.

What does that mean? Only users with the ADMIN role can enter the Query.users resolver, and only authorized users can enter the Query.me resolver, providing information about themselves. So, the role of the @auth and @hasRole directives here is to just guard the resolver, not to be entered by users with insufficient permissions.

Now, let's take a look at how this kind of directive could be implemented in TypeScript. For this example, we will use GraphQL Yoga and the Axolotl engine, as it will be used throughout the book. Here is the code of the custom directive implementation:

```
import { defaultFieldResolver } from "graphql";
import { YogaInitialContext } from "graphql-yoga";
import { createDirectives } from "@/src/axolotl.js";

export default createDirectives({
```

```
auth:(schema,getDirective) => ({
  "MapperKind.OBJECT_FIELD": (fieldConfig) => {
    const directive =
      getDirective(schema,fieldConfig,'auth')
    if(directive){
      const { resolve = defaultFieldResolver } =
        fieldConfig
      return {
        ...fieldConfig,
        resolve: async (
        source,args,context:YogaInitialContext,info) => {
          const headerValue =
            context.request.headers.get("Auth")
          if(headerValue !== 'password'){
            throw new Error("You are not authorized")
          }
          return resolve(source,args,context,info)
        }
      }
    }
  }
})
})
```

First of all, we will run this directive on just the `object` field. This is because the implementation specifies FIELD_DEFINITION, which means that this directive only works on type and interface fields.

Then, in the main part of the code – the `resolve` function – we check whether there is an Auth header with the `password` value present inside the request headers. If it is not, we throw an error. This means that all the resolvers marked with @auth directives will work only for authorized users – those that have the Authorization header with the `password` value attached to the request.

> **Note**
>
> Inside the chapter repository's folder, under `directive-auth`, you will find the full project setup with directives implemented.

Manipulating data

Directives can also be used for data manipulation or transformation. This process involves the utilization of external tools, often cloud-based services, to transform your schema based on specific rules. These rules generate a new schema that includes additional fields for other logical purposes. By doing this, we can create a backend system from the decorated schema without writing a line of code.

I will dive into more details about transformers in *Chapter 6*, but just to explain basic transformers that can be done with directives, in this example, we will use the `@model` directive to specify that our GraphQL type is a database model and the `@pk` directive to specify the primary key:

```
type Book @model {
  id: ID! @pk
  title: String!
}
```

Using these two directives, we provide our system with the transformation information it needs to generate the `Book` model and create all the database-related resolvers for it. We also specified that it should use the `id` field as a unique identifier of an object in the database.

We can also use directives inside the GraphQL query language. Some of them are `@include` or `@skip`, which allow us to conditionally include or skip fields or fragments in our queries based on runtime variables or conditions. These provide flexibility in building dynamic queries and optimizing network traffic by fetching only the necessary data.

Additionally, directives can be used for caching, rate limiting, logging, or any other custom business logic that we want to apply to the GraphQL layer. They provide a clean and declarative way to extend the functionality of our API without cluttering the schema with custom fields or arguments. In the following example, we will add cache directives to cache a book's upvotes and downvotes:

```
type Book {
  id: ID!
  title: String!
  votes: [Vote!]! @cache(maxAge: 60)
}

enum VoteKind{
  UP
  DOWN
}

type Vote {
  """
  vote value
  """
  value: VoteKind
  """
  one user has one vote per book
  """
  user: User!
}
```

Caching the upvotes and downvotes works as a good example because users don't really need this data in real time – we can afford to sacrifice up-to-date results for the sake of app speed here.

Now, let's look at the `@rateLimit` directive. Rate limit refers to the restriction placed on the number of API requests that can be made within a specified time period. It is implemented to prevent abuse, protect server resources, and ensure fair usage by limiting the number of requests a client can make within a given timeframe. The `@rateLimit` directive usually tells the backend what the rate limit should be at a certain GraphQL resolver, allowing us to create developer-facing GraphQL APIs while limiting the usage of those resolvers.

Here is a short example of the `@rateLimit` directive in use:

```
type Post {
  id: ID!
  title: String!
  content: String!
}

type Api{
  posts: [Post!]! @rateLimit(limit: 10, duration: 60)
}

type Query {
  api: Api!
}
```

By adding the `@rateLimit` directive, we limit 1 to 10 operations for every 60 seconds. With rate limits, we can easily calculate the maximum resources needed for our server, which allows us to maintain server stability.

While directives are optional, they offer a lot of flexibility and control over the behavior of our GraphQL API. By leveraging directives effectively, we can enhance the capabilities of our API, improve code reuse, and provide a more tailored experience for our clients.

Summary

In this chapter, we covered the essential principles and best practices for designing robust and efficient GraphQL schemas. We began by discussing the basics of schema design, including the use of standardized names; documenting types, fields, and arguments; implementing enums; and utilizing pagination. The rules discussed here help provide a standardized way to define our API, making it easier to understand and maintain.

Additionally, we examined the unique behavior of interfaces in GraphQL and their impact on the development environment. Custom scalars were introduced to provide better descriptions for underlying types and handle unknown dynamic types. Lastly, we briefly touched upon the practical application of directives in GraphQL, with a promise to delve deeper into them in a dedicated chapter.

In the upcoming chapter, we will delve into the intricacies of a schema writing style that can be considered a pattern, known as the "pipes pattern," and its application in various use cases within GraphQL schemas.

4

Building Pipes

One of the key challenges in building scalable and secure GraphQL APIs is ensuring that the right level of access control is implemented. By incorporating authentication information into our schemas, we can ensure that only authorized users have access to specific parts of our data graph. This not only enhances the security of our applications but also provides a seamless and intuitive experience for schema consumers, understanding what parts of schema are accessible by which kind of consumer.

In this chapter, we will dive into the concept of pipes and connections, which serve as a powerful tool for defining relationships between various components of our schema.

Pipes allow us to connect different parts of our schema, enabling us to establish relationships and dependencies between objects. This ensures that our schema is cohesive and data can flow seamlessly between different entities. We will explore various strategies for implementing pipes, which will empower us to model complex relationships in our GraphQL schemas.

Throughout this chapter, we will provide practical examples and hands-on exercises to solidify our understanding of these concepts. By the end, you will have acquired the knowledge and skills necessary to build robust and secure GraphQL schemas, enabling you to confidently design APIs that meet the specific access control needs of your applications.

In this chapter, we will cover the following topics:

- Introducing pipes
- Logical pipes
- Access pipes
- Ownership pipes

Technical requirements

To complete this chapter, you will need a GraphQL IDE such as VS Code, Sublime Text, Notepad++, or GraphQL Editor.

You can access the code in this chapter in the book's GitHub repository: `https://github.com/ PacktPublishing/GraphQL-Best-Practices/tree/main/chapter-04`.

Introducing pipes

In GraphQL, types serve as the building blocks of our schema, allowing us to define the structure and behavior of our data. One powerful aspect of types is their ability to group schema parts by the domain they represent. By organizing our schema in this way, we can achieve a clear and cohesive separation of concerns, making our code base more maintainable and easier to understand.

Pipe patterns leverage this GraphQL structure. Usually, pipes are used to pass or stop the resolver tree execution.

In addition to improving the overall organization of our schema, domain-specific types also enable us to implement fine-grained access control. By defining specific fields and permissions for each type, we can ensure that only authorized users have access to certain data or operations. This level of control is particularly crucial when dealing with sensitive information or enforcing business rules.

When we perform operations on the root query, mutation, or subscription, it does not receive any source data from another resolver. Instead, it only receives what the backend passes from the previous resolver.

However, things work differently for operations that are not executed at the root of the tree. We can pass entire objects, and importantly, this information is not inherent to GraphQL itself. It is passed during backend creation by backend developers and is visible in the framework they use.

In this example, I will show you how GraphQL is executed and how it can be used in pipes later:

```
type Person{
  friends: [Person!]!
  firstName: String!
  lastName: String!
}

schema{
  query: Query
}

type Query{
  people: [Person!]
}
```

Each `Person` object has basic information such as `firstName` and `lastName`, as well as a `friends` field to list all of the person's friends. To execute this schema, we will need two resolvers – one for `Query.people` and another for `Person.friends`. If non-root resolvers are executed in isolated environments, they need the information from the previous resolver.

In our root resolver, `Query.people`, the execution path is simple – first, it gets people from the database, and second, it returns them as an array.

But for `Person.friends`, it is not so obvious. I will illustrate how it should be executed in GraphQL using the following diagram:

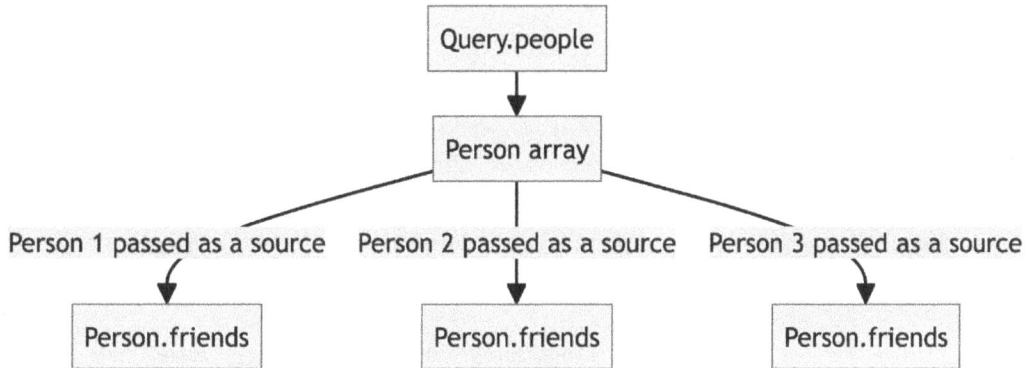

```
                          Query.people
                               |
                               v
                          Person array
            /                  |                    \
Person 1 passed as a source  Person 2 passed as a source  Person 3 passed as a source
            |                  |                    |
            v                  v                    v
      Person.friends      Person.friends       Person.friends
```

Figure 4.1: Resolution of GraphQL query

So, every `Person.friends` resolver receives a `Person` object as a source, to know how it can gather friends of that `Person`.

At this point, knowing how the GraphQL resolvers work, we can proceed to the next section about logical pipes.

Logical pipes

Logical pipes are only responsible for the logical grouping of operation types. This way, we are splitting the schema tree into a set of smaller sub-trees.

By grouping schema parts based on their domain, we can easily navigate and discuss our schema structure with others. This approach helps us maintain a modular and organized architecture, making it easier to add or modify functionality as our application evolves. We can think of these domain-specific types as containers that encapsulate related data and operations, providing a logical separation of concerns.

In the following example, I will present you with a system for reading information in a home that utilizes **Internet of Things** (**IoT**) devices, such as individual solar-charged lights connected via Wi-Fi. The division into domains will allow us to separate the specific domains that household members have access to.

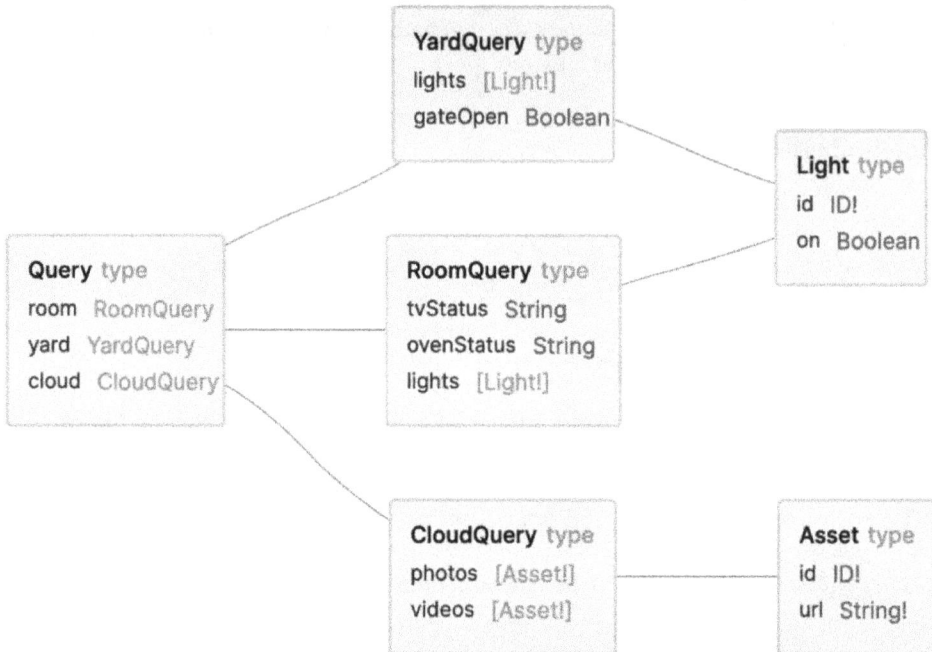

Figure 4.2: Visual diagram representing the home cloud system in GraphQL

Here, we have divided this system into three domains:

- What is happening in the room – RoomQuery

- What is happening in the backyard – YardQuery

- What shared photos and videos we have in the cloud – CloudQuery

By dividing our available queries into three types, we have organized the structure of the schema and made it easier for consumers of the schema to execute queries. Inside YardQuery, we can check what lights are on based on their unique IDs and whether the yard gate is open:

```
type YardQuery{
   lights: [Light!]
   gateOpen: Boolean
}

type Light{
   id: ID!
   on: Boolean
}
```

Inside `RoomQuery`, we can also check our lights, as well as `tvStatus` and `ovenStatus`:

```
type RoomQuery{
  tvStatus: String
  ovenStatus: String
  lights: [Light!]
}
```

Inside `CloudQuery`, we can check our family photos and videos, identifying them by ID, and get their content via a URL:

```
interface CloudQuery{
  photos: [Asset!]
  videos: [Asset!]
}

type Asset{
  id: ID!
  url: String!
}
```

As you can see, we have lights from the same manufacturer installed in both the room (`RoomQuery`) and the backyard (`YardQuery`). Dividing them into two domains will make it easier for us to check their statuses later on.

Indeed, it is worth considering when such a division is beneficial. I recommend reading more about domain-driven design, as it is a broad topic; however, put simply, **Domain-Driven Design** (**DDD**) is an approach to software development that focuses on understanding and modeling the problem domain. It advocates for organizing the code base around distinct domains or subdomains, which represent different aspects of the business or system. This division enables teams to have a clearer understanding of the different areas of the system, allows for more focused development efforts, and facilitates easier maintenance and scalability.

In the context of our example, dividing the system into domains such as "room" and "yard" makes sense because they represent distinct areas of the home with their own sets of devices and functionalities. This division helps in organizing and managing the system, making it easier to handle and reason about each domain separately.

Furthermore, we can nest one resolver within another to further break down access. For example, in this scenario, we can handle our system to control multiple houses. To achieve this, we can create another query that contains fields from the `Query` type and, in the `Query` type, differentiate houses by ID. Let's see how this looks in practice. In the following example, we will add `HouseQuery` to control every field by house:

```
type Query{
  house(id: ID): HouseQuery
```

```
}

type HouseQuery{
  room: RoomQuery
  yard: YardQuery
  cloud: CloudQuery
}
```

This way, we added another layer to our pipe types so that we can control many houses, and we did that just by moving all the fields from the root query operation to `HouseQuery` and then creating a field with its type.

Grouping schema parts by domain is a widely recognized pattern that promotes clarity, maintainability, and reusability. By utilizing domain-specific types, we can effectively structure our schema, implement fine-grained access control, and create a modular architecture that adapts to the evolving needs of our applications. That is not all, however. We can also group types by access.

Access pipes

The main reason why we use **access pipes** is to have control over which parts of the schema are available to individual groups of schema consumers. By doing this, we can see how access control will work in our system even during the graph creation phase.

To understand access pipes, we need to understand **Role-Based Access Control** (**RBAC**). RBAC is a mechanism that allows us to control access to data based on user roles. It works by assigning each user a role, which determines the resources available to them. In practice, when defining a GraphQL schema, we can specify roles for individual object types and fields. We can define different roles such as *admin*, *user*, or *guest*, and assign them specific read and write permissions for data using directives or we can do it with access pipes. When a user sends a GraphQL query, their role is taken into account during query execution. The GraphQL backend engine checks whether the user has the necessary permissions to resolve the specific type. If the user lacks the appropriate permissions, the GraphQL engine returns an appropriate "access denied" message.

This mechanism allows us to create secure and controlled API interfaces with RBAC and access pipes in GraphQL. We can ensure that only authorized individuals have access to specific types, which is particularly important for applications handling sensitive information such as user data or financial data.

It is worth noting that RBAC in GraphQL is flexible and scalable. We can easily customize roles and permissions according to our application needs. Additionally, we can also incorporate business logic when determining permissions, giving us greater control over data access.

Now, let's take a look at an example using the IoT scenario, but this time, we will restrict access to certain resources. This example will showcase access pipes. Imagine we have a family at home, and we want to restrict access to photos and videos for our children. To do this, we will split `CloudQuery` accordingly:

```
interface CloudQuery{
    photos: [Asset!]
    videos: [Asset!]
}

type KidsQuery implements CloudQuery{
    photos: [Asset!]
    videos: [Asset!]
}

type ParentsQuery implements CloudQuery{
    photos: [Asset!]
    videos: [Asset!]
}
type Query{
    parents: ParentsQuery
    kids: KidsQuery
}
```

Instead of returning photos and videos from the resolver of the `CloudQuery` type directly, we transformed `CloudQuery` into an interface and created two separate queries named `KidsQuery` and `ParentsQuery`. Then, in the backend system, we will determine who is making the request by parsing request headers and we won't allow kids to enter the `ParentsQuery` assets.

The parent and child restriction example is quite simple and clear, but we can go a step further and complicate the situation. Let's consider a different example – this time, a food delivery business where we have a restaurant, a driver, and a customer:

- The driver needs their own set of queries and mutations to be used in their application for taking orders and changing their statuses
- The restaurant needs a set of queries and mutations for placing orders and browsing order history
- The customer needs a query for checking the status of their order

Of course, there could be additional entities involved, such as a partner platform that aggregates drivers, but their set would be quite extensive, so we won't delve into that here.

First, we will look at the graph presenting the whole system and then we will go through the roles:

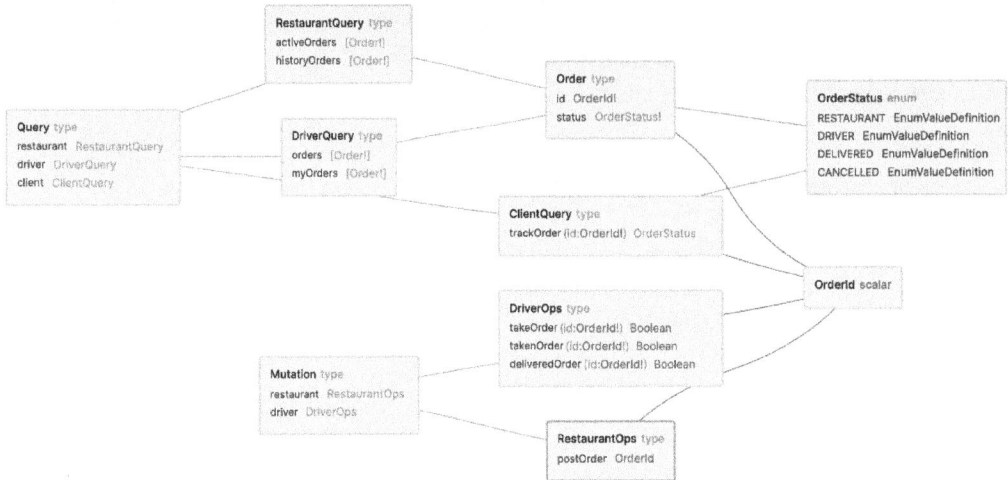

Figure 4.3: Graph presenting our delivery GraphQL schema

Within this example, we also introduce relation-based access control. **Relation-Based Access Control (ReBAC)** refers to the ability to control and restrict access to certain data based on relationships between entities. It allows you to define access rules and permissions that govern which users or roles can access specific data based on the relationships they have with other entities.

In traditional REST APIs, access control is often implemented on the server side, and the client receives a predefined set of data based on their authentication status. However, GraphQL provides a more flexible approach by allowing you to define access rules directly within the schema.

With relation-based access, you can define fine-grained rules that take into account the relationships between entities. For example, imagine you have a social media application with users, posts, and comments. You can specify that only the author of a post can edit or delete it, or only the commenter can edit or delete their own comments. These rules can be enforced at the schema level, ensuring that unauthorized users cannot perform certain operations.

As `Order` is available for both `Restaurant` and `Driver`, they can only access certain orders – `Restaurant` can only access orders that are their orders for food in their restaurant, while `Driver` can access orders that are open (nobody has taken them yet) or the active orders that are delivered by them.

We will start with the restaurant code. In our system, the customer places an order over the phone or via the internet, without using our system. Our system is only responsible for delivering the food to the customer and keeping them informed about the status. Here is how `RestaurantQuery` looks, along with its description:

```
type RestaurantOps{
    """
    inform drivers there is an order available
```

```
    """
    postOrder: OrderId
}


"""
All queries the restaurant can perform
"""
type RestaurantQuery{
    """
    orders that are not yet delivered
    """
    activeOrders: [Order!]
    """
    orders that are either delivered or cancelled
    """
    historyOrders: [Order!]
}
```

As we can see, the restaurant has one mutation, which is inserting a new order and making it available for `RestaurantOps` drivers. It can also browse active and historical orders as part of its queries.

Whoever will be implementing the application for the restaurant will have an easy task, as they don't need to know and understand the entire GraphQL schema, but only the part that is dedicated to restaurants.

Now that the restaurant has accepted the order, the driver can take the order using their mobile application, which communicates with the `Driver` part of the schema. Let's take a look at the GraphQL code snippet responsible for `Driver` operations and queries:

```
"""
All queries driver can perform
"""
type DriverQuery{
    orders: [Order!]
    myOrders: [Order!]
}

type DriverOps{
    """
    Reserves the order for the driver to be taken
    """
    takeOrder(
        id: OrderId!
    ): Boolean
    """
```

```
    Order was taken from the restaurant by the driver
    """
    takenOrder(
      id: OrderId!
    ): Boolean
    """
    order was delivered to the client
    """
    deliveredOrder(
      id: OrderId!
    ): Boolean
}
```

The driver can view pending orders using `DriverQuery.orders` and can accept orders using the `takeOrder` operation. They then proceed to pick up the order from the restaurant and, upon doing so, invoke the `takenOrder` operation and deliver the order to the customer. Once the order has been delivered, they send the `deliveredOrder` operation. The `myOrders` query displays all the orders that are taken by the driver but have not yet been delivered.

Having done the `Driver` part, let's focus on what actions the customer who ordered food from the restaurant can take. Here is a code snippet showcasing their capabilities:

```
type ClientQuery{
    """
    get the order status.
    """
    trackOrder(
      id: OrderId!
    ): OrderStatus
}

enum OrderStatus{
    """
    Order is in the restaurant
    """
    RESTAURANT
    """
    Driver has taken the order from the restaurant
    """
    DRIVER
    """
    Order is delivered to the client
    """
    DELIVERED
```

```
    """
    Order is cancelled
    """
    CANCELLED
}
```

The customer can only view the status of their order, which is represented by `OrderStatus`. We intentionally did not use the `Order` type as the `trackStatus` return type to avoid exposing fields to the customer that they should not see.

Of course, we need to attach all our operations and queries to the root `Query` and `Mutation` types so that they are visible to the consumer of our GraphQL schema:

```
type Query{
    restaurant: RestaurantQuery
    driver: DriverQuery
    client: ClientQuery
}

type Mutation{
    restaurant: RestaurantOps
    driver: DriverOps
}

schema{
    query: Query
    mutation: Mutation
}
```

Having done that, we have a basic, though fully functional, food delivery system. Here, we can see how the pipe design pattern helped us create a single schema without federation, which, despite being unified, perfectly covers the usage for each role, allowing certain users access to certain information. Understanding this pattern is also crucial for comprehending federation, which will be described in *Chapter 7*.

Access pipes are not the only possible pattern. In the next subsection, we will create pipes that also control access as well as provide ownership to the accessor.

Ownership pipes

It often happens that the owner of a specific entity should be the only one with the authority to edit it. In such cases, the **ownership pipes** pattern comes in handy, allowing editing of an object exclusively by its owner.

One of the most common use cases for ownership pipes is a blog. Here, every user of a blog is also the owner of their articles, and the only one who can edit them, add a cover image, set the publication date, and so on. Additionally, there are blog readers who can comment on each article, as well as edit their own comments and react to other comments. Once again, ownership of these objects and actions lies with the user. Furthermore, each user – whether a blog writer or commenter – has access to their own profile data and can edit it.

Just like in the previous section, we will start here by presenting the graph of our GraphQL schema for a blogging system:

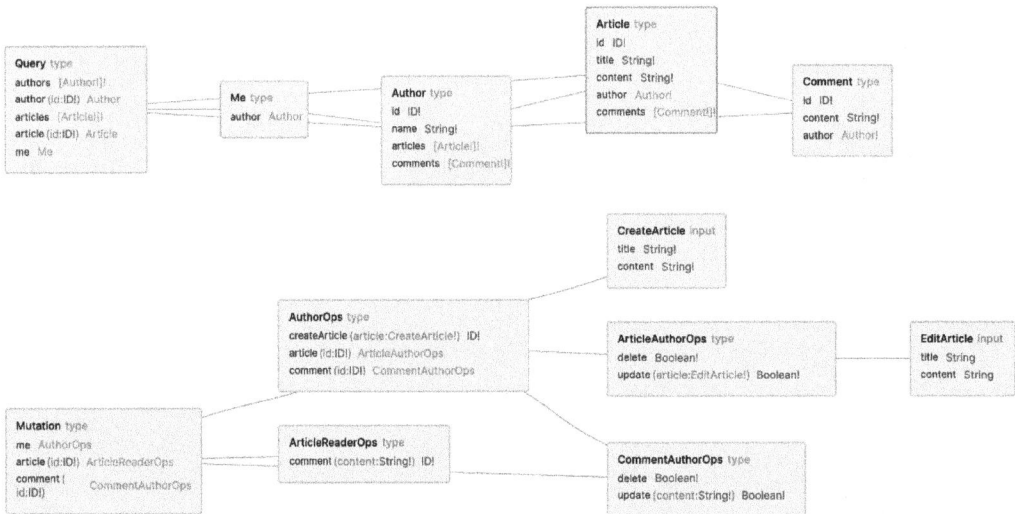

Figure 4.4: Graph presenting blog service schema

Now let's analyze the code piece by piece and explain the purpose of ownership pipes in GraphQL mutations. We'll start with the root mutation, where we have two ownership pipes, one for the administration of posts and another one for actions with other user posts:

```
type Mutation {
  """
  Operations related to the logged in blog author and its
  owned data
  """
  me: AuthorOps


  """
  Operations related to reading articles by logged in user
  """
  article(
    id: ID!
```

```
    ): ArticleReaderOps
}
```

All operations related to the author and editing their blog posts are located in the `Mutation.me` field. In the backend system, for example, based on the request header, we recognize the logged-in user and they have permission to edit objects that they own. Operations included in `Mutation.article` describe all the possible actions of the reader below any of the articles.

The difference between the two mutations is that the first (`AuthorOps`) is available only for the owner of the `Author` object while the second (`ArticleReaderOps`) can be accessed by any logged-in user. Both of them are ownership mutations; however, the article reader also owns the comments that they create below any article.

Let's take a look at the first mutation – `AuthorOps` – to see what operations are available for the schema consumer who is logged in (the blog author, in this case):

```
"""
Operations related to the logged in user
"""
type AuthorOps {
    """
    Create a new article
    """
    createArticle(
        """
        The details of the article to create
        """
        article: CreateArticle!
    ): ID!

    """
    Operations related to managing an article
    """
    article(
        """
        The ID of the article
        """
        id: ID!
    ): ArticleAuthorOps

    """
    Operations related to managing authors' comments
    """
    comment(
```

```
    """
    The ID of the comment
    """
    id: ID!
  ): CommentAuthorOps
}
```

Let's list the operations included in this code block:

- We can create articles using `AuthorOps.createArticle`
- We can edit and delete articles using `AuthorOps.article` with an ID as an argument – it does so by returning the `ArticleAuthorOps` type

Within this, `CommentAuthorOps` looks similar to the `ArticleAuthorOps` type. Its function is to combine the `delete` and `update` operation with the ID gathered from the field resolver, which means that you will be only able to use these operations after the `CommentAuthorOps` type field is executed:

```
"""
Operations related to commenting on articles
"""
type CommentAuthorOps {
  """
  Delete the comment
  """
  delete: Boolean!

  """
  Update the content of the comment
  """
  update(
    """
    The new content of the comment
    """
    content: String!
  ): Boolean!
}
"""
Operations related to managing an article
"""
type ArticleAuthorOps {
  """
  Delete the article
  """
```

```
      delete: Boolean!

      """
      Update the details of the article
      """
      update(
        """
        The updated details of the article
        """
        article: EditArticle!
      ): Boolean!
}
```

In this code example, you pass an ID to `ArticleAuthorOps` and don't have to pass it to individual `update` and `delete` mutation resolvers. Normally, we would need to pass the ID to both `delete` and `update` resolvers, and we would also need to name them differently, but this way, we increase clarity and save code lines.

The next ownership pipe is `ArticleReaderOps`. It has *reader* in its name because it only allows commenting on an article without checking the article's ownership – only checking whether the user is logged in, so they can read articles. This mutation could also hold other operations, such as liking a post and so on:

```
  """
  Operations related to reading articles
  """
  type ArticleReaderOps {
    """
    Add a comment to an article
    """
    comment(
      """
      The content of the comment
      """
      content: String!
    ): ID!
}
```

In this example, we don't own the article but the comment we made under the article, which we can delete or edit.

We can streamline the process of checking ownership in the backend by limiting it to the resolver where the owned object is accessed. Since you are already familiar with how resolvers work, you can be confident that the schema consumer's permissions have already been verified before reaching this point. This ensures that the object can only be mutated if the necessary permissions have been granted, as confirmed by a preceding pipe resolver.

This pattern not only facilitates managing the security of the schema (as, if someone can access a specific resolver, the data is automatically narrowed down to the data owned by the identified object) but it also helps us group logic and reduce the number of fields per type, thus increasing the readability of our GraphQL schema.

> **Note**
>
> At this point, we will use the Ops suffix appended to type names to further emphasize that these are collections of operations that a given consumer of the schema can perform. However, we need to remember not to mix Ops with regular types, to avoid confusion when implementing the schema by backend developers.

Summary

In this chapter, we were introduced to the first design pattern called **pipes**. This pattern is not only useful for logically dividing related types but also has many other applications.

We started with pipes that logically divide our schema, using knowledge from DDD and applying it to create a GraphQL schema. Then, we moved on to pipes that restrict access to individual query and mutation sets, allowing only authorized users to access them. Expanding upon that, we looked at ownership pipes, which connect objects and establish ownership of objects from one set to another.

In the next chapter, we will start coding and learn how to transform an existing REST API into a GraphQL one.

Part 3 - Exploring Possible Ways to Use GraphQL

In this part, we will learn how to proxy the REST backend to GraphQL. We will also write our own GraphQL transformer and learn about the concept of GraphQL Federation.

This part contains the following chapters:

- *Chapter 5, Transitioning from REST to GraphQL*
- *Chapter 6, Defining GraphQL Transformers*
- *Chapter 7, Understanding GraphQL Federation*

5

Transitioning from REST to GraphQL

In a REST API, endpoints are typically defined as specific URLs that correspond to different resources. For example, to create a new user, we would make a POST request to /users. To retrieve a list of users, we might make a GET request to the same endpoint. The server then performs the requested action or responds with the requested data.

In this chapter, we will learn how to use the existing REST API and provide a /graphql endpoint for it. We will build a proxy backend that will communicate with the REST one using Node.js and TypeScript. We will also learn about the key differences between communication in GraphQL and REST as well as similarities.

After reading this chapter, you should be able to create a GraphQL API for an existing REST backend to make the work of frontend developers and other API consumers much more pleasant.

In this chapter, we will cover the following topics:

- Why transition to GraphQL?
- Transforming a REST to-do list into GraphQL
- Passing headers
- Understanding response types
- Reviewing the disadvantages of GraphQL

Technical requirements

To complete this chapter, you will need the following:

- Terminal and Node.js LTS version installed
- {JSON} Placeholder: `https://jsonplaceholder.typicode.com/`
- Axolotl engine: `https://github.com/aexol-studio/axolotl/`
- GraphQL Yoga engine: `https://github.com/dotansimha/graphql-yoga`
- Basic knowledge of TypeScript

You can access the code in this chapter in the book's GitHub repository: `https://github.com/PacktPublishing/GraphQL-Best-Practices/tree/main/chapter-05`.

Why transition to GraphQL?

Before we get started, let's dive into GraphQL, looking at how it compares to REST and why we should use it instead of REST.

GraphQL is a query language for APIs. While GraphQL has a single endpoint, typically `/graphql`, where all queries and mutations are sent, REST relies on multiple endpoints.

With REST, we often encounter the problem of over-fetching or under-fetching data. Over-fetching occurs when the server returns more data than we need, leading to wasted resources and decreased performance. Under-fetching, on the other hand, happens when we need to make multiple requests to retrieve all the data we require, resulting in increased latency.

One of the key advantages of GraphQL is that it solves these issues by allowing the client to define the shape and structure of the response. We can specify the exact fields we want and even retrieve data from multiple resources in a single request. This flexibility empowers clients to optimize their data-fetching needs and eliminates the need for unnecessary round trips to the server.

Another advantage of GraphQL is its strong typing system. As TypeScript developers, we value static typing, and GraphQL provides a similar experience on the API side. With GraphQL, we define a schema that explicitly describes the available types, fields, and relationships in our API. This schema acts as a contract between the client and server, ensuring that both sides understand each other's expectations. This type-safety enables us to catch potential errors early in the development process and provides better tooling for developers.

Furthermore, GraphQL supports real-time updates through subscriptions. With REST, we typically rely on techniques such as long polling or WebSockets to achieve real-time functionality. GraphQL simplifies this by allowing clients to subscribe to specific data changes and receive updates in real time. This feature is especially useful for building applications that require real-time collaboration, such as chat applications or live dashboards.

In summary, while RESTful APIs have been the go-to choice for many years, GraphQL offers a more efficient and flexible approach to building APIs. With its ability to fetch only the required data, strong typing system, and support for real-time updates, GraphQL provides a compelling alternative for developers seeking a more efficient way to build and consume APIs. This is why it is a good idea to transition to GraphQL.

In the next section, we will build a REST proxy in GraphQL and TypeScript.

Transforming a REST to-do list into GraphQL

Unlike the previous examples in this book, we won't require a complicated schema here. Our goal is to create a basic to-do list using a public API. To start, we'll build a simple proxy that mimics the functionality provided by REST, without getting into the nitty-gritty details.

You might be wondering what the benefit of doing this is. Well, with GraphQL, we gain the advantage of precisely defined return types and parameters. These will be utilized by our GraphQL resolvers to direct requests to the appropriate REST endpoints. This is where the true power of GraphQL shines in comparison to REST.

Alright, let's begin with the list of available to-dos. This is the simplest example of how GraphQL enhances type safety in our to-do list project. Enter `https://jsonplaceholder.typicode.com/todos` in your browser to test the endpoint and you should receive the following response:

```
[
  {
    "userId": 1,
    "id": 1,
    "title": "delectus aut autem",
    "completed": false
  },
  {
    "userId": 1,
    "id": 2,
    "title": "quis ut nam facilis et officia qui",
    "completed": false
  },
  {
    "userId": 1,
    "id": 3,
    "title": "fugiat veniam minus",
    "completed": false
  },
  {
    "userId": 1,
```

```
    "id": 4,
    "title": "et porro tempora",
    "completed": true
  }
]
```

Now, we will define the format of the response with a GraphQL schema. This way, we will describe our REST backend with GraphQL types and achieve type-safety:

```
type Todo{
  completed: Boolean
  title: String!
  userId: Int!
  id: Int!
}

schema{
  query: Query
}

type Query{
  """
  Get list of all available todos
  """
  todos: [Todo!]!
}
```

As we can see, we have successfully typed the Todo object. Now, we need to write code that will allow us to run this REST proxy. In previous chapters, we only dealt with GraphQL, but here we need to start coding something that works, which cannot be done without using another programming language. I have chosen TypeScript here because it is one of the most popular programming languages at the time of writing this book. Of course, if you prefer, you can also try it in your preferred programming language; there are plenty of tools and servers for GraphQL available in each of them.

Let's start coding some real stuff. For our project, we will use the axolotl npm package as a wrapper for GraphQL resolvers and GraphQL Yoga for GraphQL server. To create the project, follow these steps:

1. Open a terminal and run npx @aexol/axolotl create-yoga YOUR_PROJECT. This will initiate an axolotl project with the GraphQL Yoga engine connected.

2. Next, run cd YOUR_PROJECT.

3. Then, copy the GraphQL schema defined previously in the chapter to the schema.graphql file.

4. Run `npx axolotl build`. This will generate TypeScript typings for the project, and give you the possibility to check what resolvers we can implement and what arguments the resolvers pass to our backend.

5. Create an `src/index.ts` file with the following content to implement the resolvers:

```
import { Axolotl } from '@aexol/axolotl-core';
import { Models } from '@/src/models.js';
import { graphqlYogaAdapter } from '@aexol/axolotl-
graphql-yoga';
import fetch from 'node-fetch';

const { createResolvers } =
  Axolotl(graphqlYogaAdapter)<Models>();


const resolvers = createResolvers({
  Query: {
    todos:async () => {
      const response = await fetch
        ('https://jsonplaceholder.typicode.com/todos')
      const json = await response.json()
      return json
    }
  },
});
```

Let's describe the code:

- First, we have to import the required packages. We need `axolotl-core` to specify schema and model paths.

- Then, we initiated the `Axolotl` framework with a schema path, which is, of course, our GraphQL schema and models path.

- We also need to import those models and the GraphQL Yoga adapter, making this server engine our main query processor.

- In Node.js, we also need `node-fetch` to communicate with our REST backend.

- Then, with the `createResolvers` function, we can define all the resolvers. The only one we will define now is the `Query.todos` resolver – here, we just call the REST endpoint and return its response on the resolver.

6. You can now run the server by typing the following command in your terminal:

    ```
    npm run build && npm run start
    ```

    It should run the server on `http://localhost:4000/graphql`.

7. Now, visit the server and perform the `todos` query. I am 100% sure you know how to do it but, for the sake of clarity, I will post the query here:

    ```
    {
      todos{
        completed
        title
      }
    }
    ```

    We just query all the to-dos using GraphQL and you should receive the same response we received from the REST API, but with only two fields on each object – `completed` and `title`.

Simple, isn't it? Okay, we can now go a step further and implement a GraphQL resolver for getting a to-do using its ID. It may seem complicated, but it is not impossible. Our REST endpoint for getting a to-do by its ID looks like this: `https://jsonplaceholder.typicode.com/todos/10` (in that example, `10` is the ID of the to-do).

Now, how do you transfer this functionality to GraphQL with only one endpoint, you may ask? We need to use GraphQL parameters and use them to build a URL:

1. First, we need to mirror our REST backend into the GraphQL schema. Here, we added a query to fetch a to-do by its ID:

    ```
    type Query{
      """
      Get list of all available todos
      """
      todos: [Todo!]!
      """
      Get todo by provided ID
      """
      getTodoById(
        id: Int!
      ): Todo
    }
    ```

2. Then, we need to run the `build` command again (`npx axolotl build`) for our schema changes to be reflected in our typing system. The system will provide typings for arguments passed to the resolver.

3. Now, we need to add the `getTodoById` resolver in our TypeScript code, just under the `todos` resolver. The `models` file that is generated by the Axolotl framework should automatically suggest we add this type when we add it to our `schema.graphql` file and regenerate the models. It should be put inside the `createResolvers` function, like so:

```
...
    getTodoById: async (p,args) => {
      const response = await fetch(
        `https://jsonplaceholder.typicode.com/todos/
          ${args.id}/`)
      const json = await response.json()
      return json
    }
...
```

From `args`, which is an object holding GraphQL parameters passed to the resolvers, we have the `id` field. Then, we use a template literal string to insert `args` into our REST API link and get a JSON response from it.

4. Now, visit the server and perform another query. When you run it, it should give you the object from the REST API using the GraphQL API:

```
{
  getTodoById(id:10){
    title
    completed
  }
}
```

After querying our GraphQL API, we should receive a to-do that exists in our original REST service.

At this point, we've seen how to transform simple REST calls and pass parameters. Now, we will focus on the different mutations and request methods. The first mutation that we will describe is `createTodo`. In this mutation, we need to transform simple body parameters to GraphQL input types and forward them as a `POST` method body to the REST-based backend:

```
type Mutation{
  createTodo(
    todo: TodoInput!
  ): Todo
  updateTodo(
    id: ID!
    todo: EditTodo!
  ): Todo
  deleteTodo(
    id: ID!
```

```
    ): Boolean
}

input TodoInput{
  title: String!
  userId: Int!
}

input EditTodo{
  completed: Boolean
  title: String
}
```

Here, we added inputs for our to-do mutations, as well as mutation fields on which we will create proxy resolvers.

After changing our GraphQL, remember to run the `npx axolotl models` command again to regenerate our `models` file.

Next, inside our Axolotl resolvers, then inside the `createResolvers` function, we can define the resolver as follows:

```
Mutation:{
  //...rest of resolvers
  createTodo: async (p, args) => {
    const response = await fetch(
      `https://jsonplaceholder.typicode.com/todos/`,
      {
        method:"POST",
        body:JSON.stringify(args.todo),
        headers:{
          "Content-Type":"application/json"
        }
      }
    )
    const json = await response.json()
    return json
  },
}
```

Inside a `createTodo` mutation resolver, we call the REST API with the `POST` method, including arguments in our `CreateTodo` input and serializing them using `JSON.stringify`. Then, we get the response, a `Todo` object, and return it from our resolver. Additionally, we include a `Content-Type` header to inform the REST service that we are sending the `Todo` object in JSON format.

Being able to create a to-do, we should also be able to change its state with an `updateTodo` mutation. The following is the resolver responsible for modifying the to-do, by calling the `todos` endpoint with the PUT method:

```
Mutation:{
  updateTodo: async (p, args) => {
    const response = await fetch(
      `https://jsonplaceholder.typicode.com/todos/
      ${args.id}/`,
      {
        method:"PUT",
        body:JSON.stringify(args.todo),
        headers:{
          "Content-Type":"application/json"
        }
      }
    )
      const json = await response.json()
      return json
    },
  }
```

Inside the `updateTodo` resolver, we transferred `TodoInput` into the REST PUT body. We also transferred the `id` field from arguments to the URL in the REST API.

Just to finish the REST-based GraphQL backend, I will show you how the DELETE method is implemented:

```
Mutation:{
  deleteTodo: async (p, args) => {
    const response = await fetch(
      `https://jsonplaceholder.typicode.com/todos/
      ${args.id}/`,
      {
        method:"DELETE",
      }
    )
    const json = await response.json()
    return !!json
  }
```

Here, we pass the deleted to-do ID as an argument to the URL and use the DELETE method on the REST backend.

As you may see, proxying the existing REST backend didn't take much time to finish. Basically, it is all about transferring the parameters to the URL and certain resolver fields to different methods in REST. But while we were successful in doing so, usually, the time required to write such a proxy is directly proportional to the number of endpoints in our RESTful system and their level of complexity.

In the next section, we will try to understand a couple of other aspects of proxying REST to GraphQL, starting with headers.

Passing headers

HTTP header fields are a collection of strings that are transmitted and received by both the client program and server with each HTTP request and response. Typically, these headers remain hidden from the end user and are solely handled or recorded by the server and client applications.

Since both GraphQL and REST operate based on HTTP, there isn't much difference in passing headers between them by the consumer of our system. However, header validation in REST is done at a specific endpoint, while in GraphQL, it is done at specific resolvers.

In our `todo` management REST proxy, we can simply pass those headers like this:

```
getTodoById: async (p,args) => {
  if(cachedTodosList.object){
    return (cachedTodosList.object as
      {id:number}[]).find(o => o.id === args.id)
  }
  const response = await fetch(
    `https://jsonplaceholder.typicode.com/todos/
    ${args.id}/`,
    {
      headers:{...p[2].request.headers}
    }
  )
  const json = await response.json()
  return json
}
```

We extracted the headers from the request and they are available on the third parameter of the first argument in the `graphql-yoga node` package. Then, we passed the same headers to the REST API. This way, we kept the same request headers in our REST API call that were provided to the GraphQL resolver.

As you can see, headers in GraphQL work the same way as in a REST API and you can use them the same way. Proxying them means just passing them to the next request.

In the next section, we will learn about response types.

Understanding response types

In GraphQL, one crucial aspect to keep in mind is that each resolver must provide a response of a specific type. Failure to do so will result in an error being thrown. This type of enforcement mechanism ensures that the data returned by the resolver matches the expected structure defined in the schema.

Let's examine this small schema to remind ourselves how GraphQL engines process the response:

```
type Query{
  information: Info!
}
type Info{
  lastUpdated: String!
  messages: [String!]!
}
```

In GraphQL, we can be sure of how the structure of the response will look, without entering the documentation or asking the person who created the schema. What we know from this schema is that the response structure of the `Query.information` resolver would be JSON, with the `lastUpdated` field that is for sure a `String` type, and `messages`, which is an array containing `String` instances.

What will happen if the backend developer makes a mistake and the resolver won't respond with the right fields? The backend will say that the field cannot be null. This way, we can forget about the type validation as it is provided by GraphQL itself.

In contrast, in REST, the server response can be in any shape, which is not good, especially if you are working within a team. There is no guard so the backend developer can break the contract.

Although I believe that GraphQL is much better than REST, it also has its own weaknesses compared to REST. In the next section, I will outline these weaknesses.

Reviewing the disadvantages of GraphQL

While GraphQL has gained popularity among developers for its flexibility and efficiency, it's important to acknowledge that there are some downsides to using GraphQL when compared to REST. Here are a few cons to consider:

- GraphQL introduces a new layer of complexity compared to REST. With GraphQL, developers need to define and manage a schema, which can be time-consuming and requires a deeper understanding of GraphQL concepts.

- REST is a widely adopted and understood architectural style, while GraphQL is relatively new. Developers who are already familiar with REST may need to invest time in learning GraphQL and its query language.

- Caching in GraphQL can be more challenging than in REST. Since GraphQL allows clients to specify the exact data they need, implementing caching strategies becomes more complex, as the cached data needs to be invalidated based on the specific queries made by clients.

- REST has a set of well-defined principles and standards, such as HTTP verbs and status codes, which make it easier to understand and develop APIs. GraphQL, on the other hand, is more flexible and lacks a standardized set of rules and conventions.

- While GraphQL can be more efficient in terms of reducing the number of network requests, it can also introduce performance issues if not used carefully. Poorly optimized GraphQL queries, with deep nesting and excessive data fetching, can result in increased server load and slower response times.

It's worth noting that the cons mentioned are not absolute drawbacks but rather considerations to keep in mind when deciding whether to use GraphQL or REST. The choice between the two depends on the specific requirements and constraints of the project at hand.

Summary

As we can see, even though GraphQL differs from REST, building a backend based on a functional REST service is not very complicated if you understand the service's rules.

In this chapter, we learned how to initialize a GraphQL backend in TypeScript and communicate with a REST service. Using a specific code sample, we created our backend to demonstrate the simplicity of using a REST-to-GraphQL proxy.

We also looked at the main advantages and disadvantages of REST and GraphQL, as well as discussed the similarities and differences between how their backends work by exploring topics such as caching, headers, and responses.

In the next chapter, we will learn how to build CRUD systems using a tool that was created shortly after the development of GraphQL and is not embedded in its specification – transformers.

6

Defining GraphQL Transformers

GraphQL transformers work by using directives and types from one schema to generate a new, improved schema together with backend resolvers. The principle of this solution is that each type, field, and argument can have a directive associated with it, which tells the transforming tool how to generate GraphQL code based on the information from the original schema and the directives.

Now, the original intention of this chapter was to describe several commercial solutions of directive libraries that allow us to transform GraphQL schemas. And yet, surprisingly, most of them disappeared from the market while writing this book. Does this mean that this technology is obsolete? No. Fortunately, there are still some options available.

Therefore, in this chapter, we will describe popular transformer techniques, before creating our own transformer. This will help you understand how it is possible to transform a GraphQL schema to the output schema and generated backend.

So, in this chapter, we will cover the following topics:

- Understanding how transformers work
- Creating your own transformer
- The rise and fall of transformers

Technical requirements

To complete this chapter, you will need the following:

- Terminal and Node.js installed
- A GraphQL IDE such as VS Code, Sublime Text, Notepad++, or GraphQL Editor
- A simple GraphQL parser overlay: `https://github.com/graphql-editor/graphql-js-tree`
- A basic knowledge of TypeScript

You can access the code in this chapter in the book's GitHub repository: `https://github.com/PacktPublishing/GraphQL-Best-Practices/tree/main/chapter-06`.

Understanding how transformers work

Transformers work by keeping your database model definition inside a GraphQL file. Of course, this is not our main schema file where we implement resolvers. It is a source of truth for the system that transforms the model into a generated backend, which is responsible for exposing the output schema. This usually happens by generating code for GraphQL nodes and fields that implement directives specified by the transformer technology.

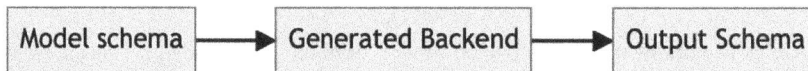

Figure 6.1: Transformer process flow

The `@model` directive is a powerful feature provided by GraphQL transformers that simplifies the process of building scalable and flexible data models for your applications. With the `@model` directive, you can easily define your data schema using GraphQL syntax, and the transformer will take care of generating the necessary code and infrastructure to manage your data.

When you apply the `@model` directive to a GraphQL type, GraphQL transformers automatically create a set of GraphQL operations and a corresponding database table to store and retrieve the data. These operations include querying, creating, updating, and deleting data, making it easy to perform **Create, Read, Update, and Delete (CRUD)** operations on your data.

The `@model` directive also allows you to define relationships between different types. For example, if you have a `User` type and a `Post` type, you can establish a one-to-many relationship between them by adding a field to the `User` type that references the `Post` type. The GraphQL transformer automatically generates the necessary resolvers and resolver templates to handle these relationships, making it easy for you to navigate and query the related data.

Let's say we have an application where users can create posts. To define our data model using the `@model` directive, we can start by creating two GraphQL types: `User` and `Post`.

First, we define our `User` type using the `@model` directive:

```
type User @model {
  id: ID!
  name: String!
  email: String!
  posts: [Post]
}
```

In this example, the User type has an id field of the ID type, a name field of the String type, and an email field of the String type. We also have a posts field, which represents the posts created by the user. This field is of the [Post] type, indicating that it is an array of Post objects.

We use the @connection directive to establish a relationship between User and Post, specifying the name "UserPosts" for this connection.

Next, let's define our Post type:

```
type Post @model {
    id: ID!
    title: String!
    content: String!
    author: User
}
```

The Post type has an id field of the ID type, a title field of the String type, a content field of the String type, and an author field of the User type.

We use the @connection directive again to establish a relationship between Post and User, specifying the name "UserPosts" to indicate that it is the same connection as in the User type.

With these definitions, the transformer automatically generates the necessary code and infrastructure to manage our data. It creates a database table for each type, along with resolvers and resolver templates to handle the CRUD operations and the relationships between User and Post.

Let's take a look at what the generated schema may look like. It will contain our original models and generated transformer code:

```
type Mutation{
    createUser(
        user: CreateUser!
    ): User
    updateUser(
        id: ID!
        user: UpdateUser!
    ): Boolean
    deleteUser(
        id: ID!
    ): Boolean
    createPost(
        post: CreatePost!
    ): Post
    updatePost(
        id: ID
        post: UpdatePost!
```

```
    ): Boolean
    deletePost(
        id: ID!
    ): Boolean
}

type Query{
    users: [User!]
    posts: [Post!]
    userById(
        id: ID!
    ): User
    postById(
        id: ID!
    ): Post
}

input CreateUser{
    name: String!
    email: String!
    postsId: [String!]
}

input CreatePost{
    title: String!
    content: String!
    userId: String!
}
input UpdatePost{
    title: String
    content: String
    userId: String
}

input UpdateUser{
    name: String!
    email: String!
    postsId: [String!]
}
```

The Mutation type contains fields that allow us to add, update, and delete each of our objects, while the Query type contains fields used to get a list of objects or to get an object by ID.

Inside the generated code, we can also see inputs generated for each of the models. Each model also contains fields that are responsible for relations. So it is possible to either specify posts when creating the User object or add a user when creating the Post object.

Now, we can easily perform operations on our data. For example, to create a new user, we can use the GraphQL mutation:

```
mutation CreateUser {
  createUser(input: {
    name: "John Doe"
    email: "john.doe@example.com"
  }) {
    id
    name
    email
  }
}
```

This mutation creates a new user with the given name and email and returns the id, name, and email values of the created user.

Similarly, we can create a new post associated with a user using the following mutation:

```
mutation CreatePost {
  createPost(input: {
    title: "My First Post"
    content: "This is the content of my first post."
    authorId: "<USER_ID>"
  }) {
    id
    title
    content
    author {
      id
      name
      email
    }
  }
}
```

In this mutation, we provide the title, content, and ID of the author (obtained from the previously created user) to create a new post. The mutation returns the id, title, and content values, and the details of the author of the created post.

By leveraging the @model directive and transformer, we can easily define and manage our data models, establish relationships between types, and perform CRUD operations, making it seamless to build applications with complex data structures.

Now, it's high time to know how it works under the hood and we can only know this by writing our own transformer.

Creating your own transformer

Transformers enable us to modify the behavior of GraphQL directives, allowing us to tailor our schema to meet our specific requirements. By writing our own transformer, we gain the power to extend GraphQL's capabilities and create a more dynamic and flexible API.

To get started, we will set up a TypeScript project and install the necessary dependencies. We will then create our custom transformer, using the provided transformer template as a starting point. We will cover important concepts such as visiting and transforming the **abstract syntax tree** (**AST**) of a GraphQL schema, and how to apply our custom logic to manipulate directives.

> **What is AST?**
>
> AST is a data structure that represents the structure of a program or code snippet. It provides a way to analyze and manipulate code programmatically. By representing code as a tree of nodes, AST allows for powerful static analysis, code transformation, and tooling capabilities in the realm of programming languages and compilers.

In our example, we will code the transformer for the @model directive that will transform our types using the @model directive into the full schema with CRUD operations.

So, grab your keyboard, and let's embark on this journey to become masters of GraphQL transformers. Together, we will unlock the true potential of GraphQL and create APIs that perfectly fit our development needs!

Setting up the project

To start, create a folder for our transformer project, then add a package.json file inside of it with the following contents:

```
{
  "name": "transformers",
  "main": "lib/index.js",
  "type": "module",
  "scripts": {
    "start": "node lib/index.js",
    "build": "tspc",
```

```
      "watch": "tspc --watch"
    },
    "dependencies": {
      "graphql": "^16.8.1",
      "graphql-js-tree": "^3.0.1",
      "graphql-yoga": "^4.0.5"
    },
    "devDependencies": {
      "@types/node": "^20.14.11",
      "@typescript-eslint/eslint-plugin": "^6.16.0",
      "@typescript-eslint/parser": "^6.16.0",
      "dotenv": "^16.4.5",
      "dotenv-cli": "^7.4.2",
      "eslint": "^8.56.0",
      "eslint-config-prettier": "^9.1.0",
      "eslint-plugin-prettier": "^5.1.2",
      "ts-patch": "^3.2.1",
      "typescript": "^5.1.3",
      "typescript-transform-paths": "^3.4.7"
    }
}
```

For this project, we need `typescript` to build it, `graphql` and `graphql-js-tree` to parse the schema and transform it to the other schema, and `graphql-yoga` as our GraphQL server. We also have scripts to build and start our application.

What is important here is the fact we also added the `module` type to the `package.json` file, telling it that our app will be `esmodule`. This is important in terms of cross-file imports as they need to use the `.js` extension inside the import path to work.

Next, for every Typescript project, we need `tsconfig.json`. Here is the file with some basic settings:

```
{
  "compilerOptions": {
    "target": "es2022",
    "module": "es2022",
    "esModuleInterop": true,
    "moduleResolution": "node",
    "incremental": true,
    "strictNullChecks": true,
    "skipLibCheck": true,
    "strict": true,
    "outDir": "./lib",
```

```
    "rootDir": "./src",
    "baseUrl": "./src",
  },
  "include": ["src/**/*"]
}
```

In this `tsconfig.json` file:

- We specified the target and module to make it `esmodule`.

- Then, we set `esModuleInterop` to `true` to change `import *` when a library/module is imported, such as `node:fs` to default imports. This is to allow default imports on those libraries.

- We specified that we want incremental compilation, to recompile the project every time we make a change.

- We get strict type-checking to avoid errors in code.

- We set `outDir` to specify where files should be outputted in `.js` format (`./lib`).

- We set up `rootDir` and `baseUrl` for relative paths, plus `include` to tell the compiler where to search for the files (in our case, inside the `./src` folder).

After setting up our project, we can move on and provide it a GraphQL Schema.

Creating the initial GraphQL schema

We now need our example schema, which we want to be transformed into a CRUD system. We can do this like so:

```
"""
Create a database model from type. Generates: list query and CRUD
operations.

Fields that implement types are treated as relations
"""
directive @model on OBJECT


"""
Defines a relationship
"""
directive @connection(fromField: String) on
FIELD_DEFINITION

type Post @model{
  content: String!
  name: String!
```

```
   author: Author! @connection
}


type Author @model{
  firstName: String!
  lastName: String!
  posts: [Post!] @connection(fromField: "author")
}
```

Here, the @model directive is responsible for annotating that the specified type is a model with its representation in the database – it will generate a database table in the database or a schema model in the document database.

Then, we have a Post type using the @model directive. The Post type has an author field, which our generation system should transform into a relation – one post can have one author. The Author type, which also uses the @model directive, has a posts field, which means that the author can have many posts.

Now, for our transformer to work, we should start by adding an id field to each model. To do so, we will use the built-in GraphQL extend type, which adds fields and directives to the original type inside the extend definition, telling the parser to add the fields during schema execution.

Transforming GraphQL with TypeScript

Now, we can start writing the actual transformation code. We will load the schema from the schema.graphql file and try to transform it to the output transformer schema, together with the generated backend:

```
import { createSchema,createYoga } from "graphql-yoga";
import fs from 'node:fs'
import path from 'node:path'
import { Parser, compileType, getTypeName, FieldType,
  Options, ParserField } from 'graphql-js-tree';
import { createServer } from 'http'

const temporaryMemory:Record<string,any[]> = {}

const transformSchema = (schema: string) => {
  const tree = Parser.parse(schema);
  const modelTypes = tree.nodes.filter((n) =>
    n.directives.find((d) => d.name === 'model')
  );
  const inputs = createModelInputs(modelTypes);
  return `
```

```
${schema}
${inputs}
type Query{
  ${modelTypes.map((mt) => `list${mt.name}s:
    [${mt.name}!]`).join('\n\t')}
}
type Mutation{
  ${modelTypes.map((mt) => `create${mt.name}(${mt.name}:
    ${mt.name}Input!): ${mt.name}`).join('\n\t')}
  ${modelTypes.map((mt) => `update${mt.name}(id:
    String!,${mt.name}: ${mt.name}UpdateInput!):
      Boolean`).join('\n\t')}
  ${modelTypes.map((mt) => `delete${mt.name}(id: String!):
    Boolean`).join('\n\t')}
}
${modelTypes.map((mt) => `extend type ${mt.name}
  { id: String! }`).join('\n')}`;
};
```

Let's look at the imported packages:

- `graphql-yoga`: This is our server runner
- `node:fs`: Used to load the schema file
- `node:path`: Used to join the path of the current working directory with the schema file path
- `graphql-js-tree`: Used to parse and transform the schema
- `http`: Used to create an HTTP server connected to the GraphQL Yoga server

Then, we created `temporaryMemory` to keep all the mutation changes in memory. So, if you make a mutation, you can see the changes reflected in the query connected to the same model as the mutation.

Next, to build a good transformer, we need to transform the schema into another GraphQL schema and also generate a backend that will operate on that schema. We will start by implementing the transformation function for our GraphQL code and then we will move to the execution code.

To begin, we first need to parse the schema. For this, we are using `Parse.parse` from `graphql-js-tree` to make nodes easier to read than if we transformed the schema into pure AST using the `graphql-js` library:

```
const tree = Parser.parse(schema)
const modelTypes = tree.nodes.filter(n =>
  n.directives.find(d => d.name === 'model'))
```

This part of the function is responsible for changing the schema string to a `ParserField` array and finding the GraphQL types that contain the model directive, so we know they will be transformed.

Next, we create additional inputs with the `createModelInputs` function:

```
const createModelInputs = (nodes: ParserField[]) => {
  const scalars = ['String', 'Int', 'Float', 'Boolean',
    'ID'];
  return nodes
    .map((mt) => {
      const args = mt.args
        .map((a) => {
          const isScalar =
            !!scalars.includes(
              getTypeName(a.type.fieldType)
            );
          if (!isScalar)
            replaceTypeWithString(a.type.fieldType);
          const compiledType =
            compileType(a.type.fieldType);
          return `${a.name}: ${compiledType}`;
        })
        .join('\n\t');
      const updateArgs = mt.args
        .map((a) => {
          if (a.type.fieldType.type === Options.required)
            a.type.fieldType = a.type.fieldType.nest;
          const isScalar =
            !!scalars.includes(
              getTypeName(a.type.fieldType));
          if (!isScalar)
            replaceTypeWithString(a.type.fieldType);
          const compiledType =
            compileType(a.type.fieldType);
          return `${a.name}: ${compiledType}`;
        })
        .join('\n\t');
      return `
input ${mt.name}Input{
  \t${args}
}
```

```
input ${mt.name}UpdateInput{
  \t${updateArgs}
}
`;
```

In this function, we create an `input` type for every transformed type. We also change the fields that are related to other types, and make the string so we can pass string IDs instead of full objects to our CRUD system.

We always need to remember relations in CRUD – if we forget about them, we will end up having nested structured objects in the database. For relations, we will use the `replaceTypeWithString` function, which replaces the relation type with the `String` type:

```
const replaceTypeWithString = (f:FieldType) => {
  if(f.type === Options.name){
    f.name = "String"
    return
  }
  replaceTypeWithString(f.nest)
}
```

Here, we are searching the nested type using a recurrent function to only change the name of the type without changing its `List` and `NonNull` features.

At the end of the `createModelInputs` function, we return inputs beginning with the name of the type and ending with the `Input` string. Every input has the same scalar fields as the original type and fields that contain relations are transformed into `String` inputs.

Next in our transform function, we generate an extension of our model types with an `id` field:

```
extend type Post { id: String! }
```

What this means is that `Post` should have an extra `id` field in it to be able to identify it in the database. Identifying models by a specific field is crucial for forming relations with other models.

Then, here is how the extended type will look like for the GraphQL server:

```
type Post @model{
  content: String!
  name: String!
  author: Author!
  id: String!
}
```

So, our `extend` type will join the field definitions of all the extensions of the `Post` type and provide one type for the GraphQL server.

Now, let's take a look at the code that generates all CRUD operations:

```
type Query{
  ${modelTypes.map(mt => `list${mt.name}s:
    [${mt.name}!]`).join('\n\t')}
}
type Mutation{
  ${modelTypes.map(mt => `create${mt.name}(${mt.name}:
    ${mt.name}Input!): ${mt.name}`).join('\n\t')}
  ${modelTypes.map(mt => `update${mt.name}
    (id: String!,${mt.name}: ${mt.name}Input!):
      Boolean`).join('\n\t')}
  ${modelTypes.map(mt => `delete${mt.name}(id: String!):
    Boolean`).join('\n\t')}
}
```

Here, we added a query to list all the objects in our system, plus three mutations to manipulate the model data:

- `create` to create an object with our prepared input
- `delete` to remove an object with the `id` field we added using `extend`
- `update` that uses both of them to update the object

The result of the transform function is the newly generated schema.

Now, what's left is writing the backend generator to implement the generated resolvers, like so:

```
type Post @model{
  content: String!
  name: String!
  author: Author! @connection
}

type Author @model{
  firstName: String!
  lastName: String!
  posts: [Post!] @connection(fromField: "author")
}


input PostInput{
  content: String!
  name: String!
  author: String!
```

```
}
input PostUpdateInput{
  content: String
  name: String
  author: String
}


input AuthorInput{
  firstName: String!
  lastName: String!
  posts: [String!]
}
input AuthorUpdateInput{
  firstName: String
  lastName: String
  posts: [String!]
}

type Query{
  listPosts: [Post!]
  listAuthors: [Author!]
}
type Mutation{
  createPost(
    Post: PostInput!
  ): Post
  createAuthor(
    Author: AuthorInput!
  ): Author
  updatePost(
    id: String!,
    Post: PostUpdateInput!
  ): Boolean
  updateAuthor(
    id: String!,
    Author: AuthorUpdateInput!
  ): Boolean
  deletePost(
    id: String!
  ): Boolean
  deleteAuthor(
    id: String!
```

```
    ): Boolean
  }
  extend type Post { id: String! }
  extend type Author { id: String! }
```

This is the code that is generated by our transformer library. What is happening here?

Our transformer generated `id` fields for the models. It did so because we need to identify those objects in the database.

Then, it generated inputs as they are needed by the `create` and `update` functions. Those inputs contain the same scalar fields as the types they were generated from. This allows us to use the output schema to insert data into the database.

The generated query resolvers consist of list functions for our available models. Those list functions return a list of all objects of a certain type from the database. The generated mutation gives you access to all CRUD mutations with an `id` field as an accessor to the object.

However, this is not the end – we need to write resolvers and run the GraphQL server now. To do this, we will use `yoga` and implement our dynamic per-model resolvers inside the `resolvers` field. The dynamic per-model resolvers will generate the backend for each type, with the @`model` directive inside the original GraphQL schema.

Let's see how we do this. I will present the code part by part, but you can see the whole example together in the chapter's repository code. Here's the first part:

```
const run = async () => {
  const schemaFile = fs.readFileSync(
    path.join(process.cwd(), './schema.graphql'),
    { encoding: 'utf-8', }
  );
  const newSchema = transformSchema(schemaFile);
  const tree = Parser.parse(schemaFile);
  const modelTypes = tree.nodes.filter((n) =>
    n.directives.find((d) => d.name === 'model')
  );
```

Using `fs.readFileSync`, we loaded the schema file as a string, then used this string to pass it to our transform function that generates a new schema. After that, using the `Parser.parse` function, we parsed the schema to use the `ParserField` type nodes for schema composition. We filter out only those nodes that are implementing the @`model` directive in our source schema.

Next, we will start specifying resolvers from the fields that implement the @`connection` directive:

```
const connectionFunction = (a: ParserField) => {
  return [
```

```
    a.name,
    (source: any) => {
      const relatedObjectFieldName = a.directives
        .find((d) => d.name === 'connection')
        ?.args.find((a) => a.name ===
          'fromField')?.value?.value;

      const argTypeName = getTypeName(a.type.fieldType);
      const isArrayField =
        a.type.fieldType.type === Options.array ||
        (a.type.fieldType.type === Options.required &&
          a.type.fieldType.nest.type === Options.array);
      if (relatedObjectFieldName) {
        if (isArrayField) {
          return temporaryMemory[argTypeName]
          .filter((ta) =>
{
            const fieldInRelatedObject =
              ta[relatedObjectFieldName];
            if (Array.isArray(fieldInRelatedObject))
              return fieldInRelatedObject
                .includes(source.id);
            return fieldInRelatedObject === source.id;
          });
        }
        return temporaryMemory[argTypeName].find((ta) => {
          const fieldInRelatedObject =
            ta[relatedObjectFieldName];
          if (Array.isArray(fieldInRelatedObject))
            return fieldInRelatedObject
              .includes(source.id);
          return fieldInRelatedObject === source.id;
        });
      }
      if (isArrayField) {
        return temporaryMemory[argTypeName].filter((ta) =>
          source[a.name].includes(ta.id));
      }
      return temporaryMemory[argTypeName][source[a.name]];
    },
  ] as const;
};
```

Inside the connection function, we start by checking the fromField name from our connection directive. Depending on whether such a field is specified or not, we check whether the type of the field is Array. Finally, we return the objects from temporary memory.

Now, we will use the connection function on all the nodes that have the @connection directive implemented:

```
const relations = Object.fromEntries(
  tree.nodes
    .filter((n) => n.args.some((a) =>
      a.directives.find((d) => d.name === 'connection')))
    .map((n) => {
      return [
        n.name,
        Object.fromEntries(
          n.args.filter((a) => a.directives.find((d) =>
            d.name === 'connection'))
            .map(connectionFunction),
        ),
      ] as const;
    }),
);
```

Next, we need to generate list resolvers for our @model types. The code responsible for generating resolvers is dynamic and based on all the types that have the @model directive implemented. So, we need to iterate those types and map them to the actual resolver functions:

```
const generateListResolvers = (modelTypes: ParserField[])
=> {
  return Object.fromEntries(
    modelTypes.map((mt) => [
      `list${mt.name}s`,
      () => {
        return temporaryMemory[mt.name];
      },
    ]),
  );
};
```

In this code, we iterate all the model types and provide a query for them. For example, for the Post model, the name of the Query field will be listPosts, and for the Author model, the name will be listAuthors, and so on. So, every model receives a set of queries and mutations that allow full interaction with its database objects.

Then, our second array element is the actual resolver function that returns a list of objects from our temporary data store. After that, the built-in `Object.fromEntries` function converts arrays such as `[[k1,v1], [k2,v2]...]` into objects such as `{ k1: v1, k2:v2 }`.

Let's also analyze one mutation to understand how those objects get into the temporary store:

```
const generateMutationResolvers = (modelTypes:
ParserField[]) => {
  return {
    ...Object.fromEntries(
      modelTypes.map((mt) => [
        `create${mt.name}`,
        (_: any, args: any) => {
          temporaryMemory[mt.name] ||= [];
          const creationPayload = {
            ...args[mt.name],
            id: '' + idCounter++,
          };
          temporaryMemory[mt.name].push(creationPayload);
          return creationPayload;
        },
      ]),
    ),
    ...Object.fromEntries(
      modelTypes.map((mt) => [
        `delete${mt.name}`,
        (_: any, args: any) => {
          temporaryMemory[mt.name] ||= [];
          if (!temporaryMemory[mt.name].find((o) =>
            o.id === args.id)) return false;
          temporaryMemory[mt.name] =
            temporaryMemory[mt.name]
              .filter((o) => o.id !== args.id);
          return true;
        },
      ]),
    ),
    ...Object.fromEntries(
      modelTypes.map((mt) => [
        `update${mt.name}`,
        (_: any, args: any) => {
          temporaryMemory[mt.name] ||= [];
          if (!temporaryMemory[mt.name].find((o) =>
            o.id === args.id)) return false;
```

```
            temporaryMemory[mt.name] =
              temporaryMemory[mt.name].map((o) =>
                o.id !== args.id ? o : { ...o,
                  ...args[mt.name] },
            );
            return true;
          },
        ]),
      ),
    };
};
```

In this generated mutation, fields have names based on types with the attached @model directive. The first line of the resolver checks whether there is an array in the store for the name of the model. Then, we push the object from the parameter named after the object, making it a universal object store that relies on a GraphQL type name. This way, we store the objects in our temporary store.

Now, we can use all our functions and insert those resolvers into the executable schema:

```
const schema = createSchema({
  typeDefs: newSchema,
  resolvers: {
    ...relations,
    Query: generateListResolvers(modelTypes),
    Mutation: generateMutationResolvers(modelTypes),
  },
});
const yoga = createYoga({ schema });
const server = createServer(yoga);
server.listen(4000, () => {
  console.info('Server is running on
    http://localhost:4000/graphql');
});
}
run()
```

We can also query the resolvers if our temporary backend is running. Let's see that next.

Running the transformed backend

Now that we have the generated schema and generated resolvers, we are ready to start our Yoga server and use GraphiQL to play with some real data. Run the following command:

```
npm run start
```

> **Note**
>
> If you haven't built your backend yet, run `npm run build` to start it.

Now, you should see the server running on port 4000. When you enter `http://localhost:4000/graphql` inside the browser, you should then see the GraphiQL console where you can execute queries and mutations on your generated backend.

To finish up, here are a few ideas for directives you can add to your transformer solution to make it more complete:

- `@auth`: A directive used for authorization to define roles of users who are eligible to see certain data
- `@object`: A directive used to inform your transformer system that the type should not be stored in the database as a separate model, but as an object stored inside another type field
- `@richText`: A directive used to inform the user that the `String` field contains rich text instead of pure String
- `@rest`: A directive used to obtain data from the REST endpoint

You can also modify the schema by adding more models.

To finish off this chapter, in the next section, we will briefly discuss the role that transformers played in promoting the GraphQL technology.

The rise and fall of transformers

In this rather short section, I will try to explain why most transformer technologies are gone these days.

Transformers were a pattern that created gravity around the GraphQL ecosystem – they showed people that with GraphQL, you could define your system and execute it. Giving users the technology to write the schema, attach directives, and have the backend with the database and authorization ready was a really good feature that made GraphQL popular.

But while it was advertised that with the use of transformers, you could create your backend with GraphQL only, this was not without drawbacks. If the backend was just for really simple database operations, then you could make it with transformers, but when complexity was needed, users were forced to migrate to another technology – repelling them from transformers.

Unfortunately, some transformers failed to meet the requirements of complex use cases. They often lacked the necessary flexibility to handle dynamic scenarios where the data structure might change or where additional data transformations are required. This rigidity can lead to suboptimal solutions, requiring developers to resort to manual data manipulation or custom code.

To overcome these limitations, it is essential to choose transformers that provide a high degree of flexibility. A flexible transformer allows us to easily adapt to evolving data requirements, seamlessly accommodating changes without sacrificing performance or introducing complexity. This flexibility empowers developers to efficiently transform data in a way that aligns with their specific needs, ultimately enhancing the overall GraphQL experience. That's why it is important to know how to write your own transformers, and why I taught you how to do so in this chapter.

Summary

In this chapter, we started by learning how transformers work, including the use of the transformer @model directive. Then, we wrote our own transformer system that provides a working backend for our transformed schema. Finally, based on the decline of external transformer technologies, I placed a hypothesis that to make the most of transformers, you should have your own transformer system under your own control.

In the next chapter, we will learn about another GraphQL concept that needs external tools to work: federation. We will learn how to effectively connect multiple GraphQL-based systems and join them as one big schema.

7

Understanding GraphQL Federation

In the world of GraphQL, one of the most powerful features is the ability to include one schema within another. This allows us to create modular and reusable components, making our overall schema more organized and maintainable.

With GraphQL Federation, you can create a "gateway" that acts as the single entry point for all client requests. This gateway is responsible for routing the requests to the appropriate microservices, combining the responses, and returning them to the client as a single unified GraphQL response. This way, we can host one backend service that combines and proxies our requests to other GraphQL servers.

More widely, GraphQL Federation simplifies the process of building large-scale GraphQL APIs by breaking them down into smaller microservices, while still providing a unified and seamless experience for your clients. With GraphQL Federation, you can build scalable, maintainable, and extensible GraphQL APIs that meet the evolving needs of your application.

In this chapter, you will learn why you need federation inside GraphQL, as well as the basic rules of federation.

So, in this chapter, we will cover the following topics:

- Why do we need federation?
- Understanding the basic rules of federation
- Merging schemas for federation
- Federating schemas in a to-do list service
- Meeting the need for GraphQL Federation

> **Note**
>
> Remember that federation is more of a conceptual idea right now and there are currently many different specifications for federation. Here, I am just presenting the general concepts of federation, but they may differ across different implementations.

Technical requirements

To complete this chapter, you will need a GraphQL IDE such as VS Code, Sublime Text, Notepad++, or GraphQL Editor.

You can access the code in this chapter in the book's GitHub repository: `https://github.com/PacktPublishing/GraphQL-Best-Practices/tree/main/chapter-07`.

Why do we need federation?

As established in the introduction, **GraphQL Federation** is an incredibly powerful tool for building scalable and flexible GraphQL APIs. It allows you to break down a monolithic GraphQL schema into smaller, more manageable services called **microservices**. Each microservice represents a specific domain or functionality within your application.

The possibilities for federated microservices are endless, but nevertheless, here are a few:

- Your backend can integrate with external GraphQL services, even if you don't have access to the source code

- Your schema is really big and you want to break it into smaller, more manageable schemas

- You are using multiple REST and GraphQL services but want to provide a unified GraphQL server experience

By including one schema within another, we can leverage the benefits of modular design and code reuse. It allows us to create smaller, more focused schemas that can be combined to form a larger and more comprehensive API. This approach also promotes the separation of concerns, making it easier to maintain and evolve our GraphQL infrastructure.

This infrastructure can be decentralized with the use of federation. Federation differs from other programming languages because it needs a network to work between the schemas that compose the **supergraph** (the consumer-facing schema composed of the different GraphQL API services).

The following diagram shows how federation works:

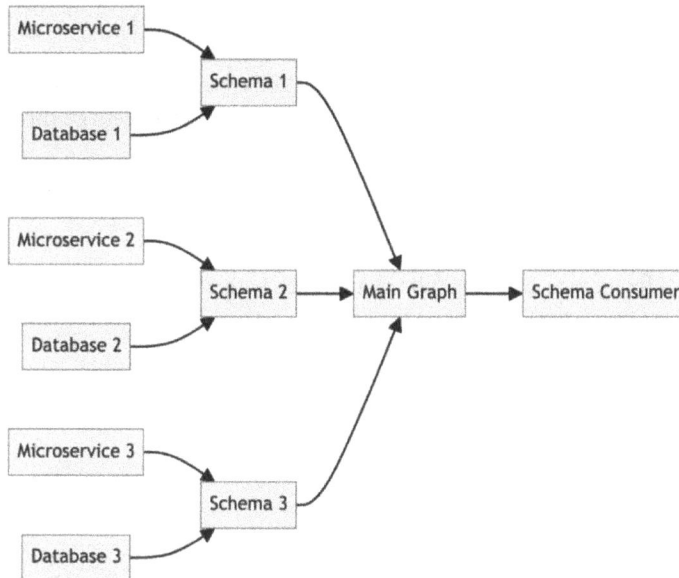

Figure 7.1: Federation graph

In this diagram, you can see that the information about the different schemas and services is hidden from the schema's consumer. In the end, the schema consumer is faced with one big schema instead of connecting to different services.

The reason we need federation here is to isolate different parts of the projects. From that, we will gain many benefits:

- Different team members can work on smaller individual parts of the bigger service. This is because, when you build a subgraph, you only need to follow the federation rules to avoid conflicts. All of this makes working on the service safer and more robust.

- By utilizing different architectures, we have the flexibility to employ various programming languages when working with subgraphs. This means that different teams can work on different subgraphs using their preferred programming languages. This approach allows for greater diversity and specialization within the development process, enabling teams to leverage their expertise and use the most suitable language for their specific subgraph using GraphQL as a glue for different microservices.

- If one microservice is faulty, the others will still be functional.

- One way to simplify the setup of different endpoints and separate authorizations is by providing a single point of access to every microservice for schema consumers. This not only eliminates the need for complicated configurations but also streamlines the access process for users.

Now that you know why you may need federation, in the next section, we will explore its basic rules.

Understanding the basic rules of federation

At the time of writing this book, there are no official federation specifications; however, there are some rules that are common across the different federation solutions available. Let's go through these rules in a question-and-answer format.

How do you merge different types?

Merging types with different type names results in a schema with all the types. So, let's say that in one schema you have an `Animal` type:

```
type Animal{
  name: String!
}
```

And in the other schema, you have a `Car` type:

```
type Car {
  name: String!
}
```

Then the resulting merged schema will include both:

```
type Animal{
  name: String!
}
type Car {
  name: String!
}
```

To simulate how schemas are merged with the federation process, we will use a small script from the GitHub chapter repository. Follow these steps:

1. Clone the book repository and enter the `chapter-07` folder inside your IDE command line.

2. Next, run the `npm i` command to install the dependencies defined in `package.json`.

3. Then, run the `npm start` command. This command will traverse the `schemas` folder and try to merge all code from schemas located in every `input` folder and output it to the `output` folder.

Here is the resulting script content in `./index.ts`:

```
import { mergeSDLs } from 'graphql-js-tree';
import fs from 'node:fs'
import path from 'node:path'
```

```
const mergeSchemas = (examplePath:string) => {
    const inputPath = path.join(examplePath,'input')
    const outputPath = path.join(examplePath,'output')
    const inputSchemas = fs.readdirSync(inputPath)
        .filter(f => f.endsWith("graphql"))
        .map(f =>
            fs.readFileSync(
                path.join(inputPath,f),
                'utf-8'
            )
        )
    let baseSchema = ``
    for( const schema of inputSchemas){
        const merge = mergeSDLs(baseSchema,schema)
        if(merge.__typename === 'error'){
            throw new Error(
                `[${inputPath}], Cant merge the schemas.
                ${merge.errors.map(e => `Conflict on:
                    ${e.conflictingNode}.
                    ${e.conflictingField}`
                )}`
            )
        }
        baseSchema = merge.sdl
    }
    console.log(`Merge successful`, outputPath)
    fs.writeFileSync(
        path.join(outputPath,'schema.graphql'),
        baseSchema
    )
}
```

Our `mergeSchemas` function loads the files from the `input` and `output` folders of an example path. Next, it finds all the files with the `.graphql` extension inside the example `input` folder. Then, file by file, it tries to merge schemas. In the end, it writes the results of the merge to the example `output` folder.

Now, we need to run this function on every example inside the repository:

```
const mergeAll = () => {
    const examplePath = path.join(process.cwd(),'schemas');
    const examples = fs.readdirSync(examplePath)
    for(const example of examples){
        try {
```

```
                    mergeSchemas(path.join(examplePath,example))
            } catch (error) {
                if(error instanceof Error)
                    console.error(error.message)
            }
        }
    }
}

mergeAll()
```

Here, we used the `mergeAll` function. This function lists all the examples from the `schemas` folder inside the repository and runs the `mergeSchemas` function on each folder.

What if different types have the same name?

Sometimes, we have types with the same names in different schemas. For example, different schemas can have different authorization systems and different user data, and by using federation, we can merge them into one big `User` type.

In the following example, we will merge schemas containing a `Car` type – one schema is a car encyclopedia schema and another is a car shop schema. Using federation, we can hold just `id` in both schemas and we don't have to keep the detailed `Car` data inside the shop schema.

So, our first schema holds the information about cars – including their make, model, and year of production – and uses the `id` field as a primary key:

```
type Car {
  id: ID!
  make: String!
  model: String!
  year: Int!
}
```

Then, our second schema holds only information about car prices:

```
type Car {
  id: ID!
  price: Float!
}
```

Here is the result of merging those two schemas:

```
type Car {
  id: ID!
```

```
    make: String!
    model: String!
    year: Int!
    price: Float!
}
```

In this example, fields from both schemas are merged into one type. This is possible because we have different field names. However, there is one exception – the id field – which has exactly the same type and is shared between two schemas. This allows us to identify the same car across different GraphQL services.

This is an ideal scenario, but what happens if we provide fields with the same names on different subgraphs? In this first schema, the Car type has a price parameter with the Float type:

```
type Car {
    id: ID!
    price: Float!
}
```

But in the second schema, the Car type has a price parameter with the Int type:

```
type Car {
    id: ID!
    price: Int!
}
```

This will obviously not be composed as the types of the price fields are different. It is not possible to merge types with the same fields of different types as we don't have a mechanism to provide information about the field origin to the schema consumer.

How do you merge field arguments?

Merging field arguments doesn't work the same as merging type fields. Let's see an example.

In this example, we will need to fetch the Car object from different databases about different car brands. In one brand database, color is the field argument used in the schema, and in the other, it's topSpeed. The reason we need to have optional arguments inside the schemas, even if we don't use them inside that service, is for GraphQL Federation to work with GraphQL validation.

In our first schema, we can get similar cars to the Car object by using topSpeed and color:

```
type Car {
    id: ID!
    getSimilar(color:String, topSpeed: Int!): [Car]
}
```

If this schema receives an argument that it doesn't recognize, it will simply throw an error.

In our second schema, we want to get cars with similar colors. Just like in the first schema, we need the topSpeed argument here so as not to break the GraphQL execution:

```
type Car {
  id: ID!
  getSimilar(color: String!, topSpeed:Int): [Car]
}
```

Now, here is the result of merging those two schemas:

```
type Car {
  id: ID!
  getSimilar(color: String!, topSpeed:Int!): [Car]
}
```

As both parameters are required in their schemas, our composed schema needs to make both parameters required for the server to work. Remember that by using federation, we will call every schema implementing the function.

> **Note**
>
> A key distinction between merging fields and arguments lies in the way they combine data. Merging fields employ a union operation, while arguments employ an intersection operation. In set theory, the **intersection** operation refers to the process of finding the common elements between two or more sets. It returns a new set that contains only the elements that exist in all of the given sets, allowing us to identify the shared elements and analyze their properties.

How do you merge different subgraphs that have different entities?

We may be in a situation where a type is returned from a field that doesn't exist in another subgraph. At the same time, on the returned type, we have merged fields from different schemas. In this case, we need to tell our federation system how to fetch the entity from a different subgraph if it is not returned by a contained field.

To get the same object from a different subgraph, we use primary key fields that identify the same object in a different subgraph, for example, the id field of an object.

In addition, it is crucial to provide instructions to a federation system on how to retrieve data from other schemas when merging the returned objects. This can be achieved by implementing the getObjectById method on each type that is federated. This allows the federation system to effectively merge the types.

Now that we have seen the basics of federation, let's create a more sophisticated example. In the next section, we will create a supergraph out of three related schemas.

Merging schemas for federation

In this example, we will take three user-related schemas and merge them into one schema. Joining them together will allow the schema consumer to use one system to manage three independent microservices that were developed separately. This process is illustrated here:

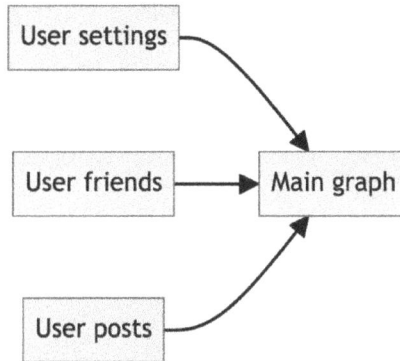

Figure 7.2: User schema composition

The first schema manages the user's profile, the second schema handles the user's friends, and the third schema is responsible for creating and reading posts on our portal.

First, let's create the schema for managing the user's profile:

```
type User @key(fields: "id"){
  id: ID!
  firstName: String!
  lastName: String!
}

type Query{
  me: User!
  getUserById(
  id: ID!): User
}

type UserOps{
  setProfile(userInput: UserInput!): User
}

type Mutation{
  me: UserOps
}
```

```
input UserInput{
  firstName: String!
  lastName: String!
}
```

We've added a query that returns the current user's details and a `Mutation` field called `me` that allows for the modification of these details. There is also a `@key` directive that allows us to determine the primary key of each object and stitch objects from multiple subgraphs together.

Our next task is to create the second schema, which manages the user's friends:

```
type User @key(fields: "id"){
  friends: [User!]
  id: ID!
}

type Query{
  me: User!
  getUserById(id: ID!): User
}

type UserOps{
  addFriend(userId: ID!): User
  removeFriend(userId: ID!): Boolean
}

type Mutation{
  me: UserOps
}
```

Notice that we have the `User` type and the same fields on `Query` and `Mutation` that we already have inside our first schema.

It is crucial to possess this knowledge in order to prevent the occurrence of different field names referring to the same operation within federated schemas. This could lead to the user updating a field in one schema while leaving it untouched in another, which would mean losing the biggest benefit of federation and having to call multiple operations to update one object.

Our third schema is responsible for posting content by each user and the possibility of reading users' posts. Here it is:

```
type User @key(fields: "id"){
  id: ID!
  posts: [Post!]
}
```

```
type Query{
  me: User!
  getUserById(id: ID!): User
}

type UserOps{
  post(post: PostInput!): Post
}


type Mutation{
  me: UserOps
}

type Post{
  id: ID!
  title: String!
  content: String!
  author: User!
}

input PostInput{
  title: String!
  content: String!
}
```

In this last schema, we allow the user to post content using the `UserOps.post` mutation. We also added the `posts` field to the `User` type, which returns the user's posts.

Now that we have three schemas in place, we need to create a schema that joins them together. This is the result:

```
type Post @key(fields:"id") {
  id: ID!
  title: String!
  content: String!
  author: User!
}

input PostInput {
  title: String!
  content: String!
}
```

```
type User @key(fields: "id") {
  friends: [User!]
  id: ID!
  posts: [Post!]
  firstName: String!
  lastName: String!
}

type Query {
  me: User!
  getUserById(id:ID!): User
}

type UserOps {
  addFriend(userId: ID!): User
  removeFriend(userId: ID!): Boolean
  post(post: PostInput!): Post
  setProfile(userInput: UserInput!): User
}

type Mutation {
  me: UserOps
}

input UserInput {
  firstName: String!
  lastName: String!
}
schema{
  query: Query,
  mutation: Mutation
}
```

Upon closer examination, we can observe that the three schemas have been consolidated into a single supergraph. Nodes that were previously exclusive to a specific schema now coexist within the supergraph. Additionally, nodes with identical names have been merged, resulting in a unified entry point for schema consumers.

You should now understand the concept of federation. In the next section, we will create our own federated service.

Federating schemas in a to-do list service

In this section, we will create a to-do list using GraphQL Federation. We will use the `axolotl` library, together with its micro-federation starter. To bootstrap the federated project, use the following command:

```
npx @aexol/axolotl@latest create-federation-yoga federated-backend
```

This will create the `federated-backend` folder and install the required dependencies inside it. It also contains a schema for a to-do list with simple user authentication and authorization.

Now, we will examine the content of the repository. We will start with the to-do list, and then move on to the users schema. In the end, we will see how the schemas merge into a supergraph and one gateway for the resolvers.

To-do list

In the to-do list schema – located inside `src/todos/schema.graphql` – we have the to-do list logic. Let's take a look at the code:

```
type Todo {
  _id: String!
  content: String!
  done: Boolean
}
type User {
  _id: String!
}

type TodoOps {
  markDone: Boolean
}

type AuthorizedUserMutation {
  createTodo(content: String!): String!
  todoOps(_id: String!): TodoOps!
}
type AuthorizedUserQuery {
  todos: [Todo!]
  todo(_id: String!): Todo!
}
```

As you can see, we have the `Todo` type that holds the content of a `Todo` object, its unique `_id` field of the `String!` type, and a `done` field that tells us whether the to-do is done.

Then, we have the `User` type, which only has the `_id` field. We also have a `TodoOps` type for marking the to-do as done.

Finally, inside `AuthorizedUserMutation`, we have the option to create a to-do or execute `todoOps`, and inside `AuthorizedUserQuery`, we have the option to list the to-dos or view one to-do by its `_id` value.

Now we know the to-do schema, we can take a look at the `src/todos/resolvers.ts` code to see how the resolvers for this schema are implemented. Here is the content of this file:

```typescript
import { createResolvers } from '@/src/todos/axolotl.js';
import { TodoModel, db } from '@/src/todos/db.js';
import { User } from '@/src/todos/models.js';

export default createResolvers({
  AuthorizedUserQuery: {
    todos: async ([source]) => {
      const src = source as User;
      return db.todos
        .filter((todo) => todo.owner === src._id);
    },
    todo: async ([source], { _id }) => {
      const src = source as User;
      return db.todos.find(
        (todo) =>
          todo._id === _id &&
          todo.owner === src._id
      );
    },
  },
  AuthorizedUserMutation: {
    createTodo: async ([source], { content }) => {
      const src = source as User;
      const _id = Math.random().toString(8);
      db.todos.push({ owner: src._id, content, _id });
      return _id;
    },
    todoOps: async ([source], { _id }) => {
      const src = source as User;
      return db.todos.find(
        (todo) =>
          todo._id === _id &&
          todo.owner === src._id
      );
```

```
      },
    },
    TodoOps: {
      markDone: async ([source]) => {
        const src = source as TodoModel;
        const index = db.todos.findIndex(
          (todo) => todo._id === src._id
        );
        db.todos.splice(index, 1, {
          ...src,
          done: true,
        });
      },
    },
  });
```

The `AuthorizedUserQuery.todos` resolver not only displays a list of to-dos based on the owner's ID (`_id`) but also checks the ownership of a `Todo` object.

Then, our `AuthorizedUserMutation.createTodo` allows us to create a to-do by adding the `owner` field to the object. It finds the to-do, again, by its `_id` value and ownership, and passes it to the next resolver.

Finally, `TodoOps.markDone` is being used to change the `Todo` object status to `done`.

Users schema

As you may have noticed already, there are no root query and mutation types in our schema and we are getting the user's `_id` field from our source. However, there are no dedicated resolvers. They do exist but in the other schema, though. Let's take a look at `src/users/schema.graphql`:

```
type User {
  _id: String!
  username: String!
}

type Mutation{
  user: AuthorizedUserMutation!
  login(username:String!, password:String!): String!
  register(username:String!, password: String!): String!
}

type Query{
  user: AuthorizedUserQuery!
}
```

```
type AuthorizedUserMutation{
  changePassword(newPassword:String!): Boolean
}
type AuthorizedUserQuery{
  me: User!
}
```

Our schema starts with a full `User` type with `_id` and `username` fields. In our to-do schema, we had the same `User` type but only with the `_id` field, so both schemas will merge because the type of the `_id` field is the same across all the schemas.

Then, we have a `Mutation` type with a pipe to check whether the user is logged in on the `Mutation.user` field, as well as two public resolvers for login and register purposes.

Next, we have the `Query` type with a `user` field that checks whether the user is logged in.

Finally, we have two types: `AuthorizedUserMutation` with a resolver to change a password, and `AuthorizedUserQuery` with a me field to retrieve the full `User` object.

After understanding the schema, we can move on and take a look at the resolvers. Inside `src/users/resolvers.ts`, there is really simple authorization and authentication logic implemented. Here is the code:

```
import { UserModel, db } from '@/src/users/db.js';
import { createResolvers } from '@/src/users/axolotl.js';

export default createResolvers({
  AuthorizedUserQuery: {
    me: ([source]) => {
      const src = source as UserModel;
      return src;
    },
  },
```

At the beginning of the file, you can see the `AuthorizedQuery.me` resolver, which, just like in the to-do schema resolvers, expects the `User` object to be passed.

Continuing on in the file, let's take a look at the authentication and authorization resolvers:

```
  Mutation: {
    login: async (_, { password, username }) => {
      return db.users.find(
        (u) => u.username === username &&
```

```
              u.password === password
      )?.token;
    },
    register: async (_, { password, username }) => {
      const userExists = db.users.find(
        (u) => u.username === username
      );
      if (userExists)
        throw new Error(
          'User with that username already exists'
        );
      const token = Math.random().toString(16);
      const _id = Math.random().toString(8);
      db.users.push({
        _id,
        token,
        password,
        username,
      });
      return token;
    },
    user: async (input) => {
      const token = input[2].request.headers.get('token');
      if (!token) throw new Error('Not authorized');
      const user = db.users.find((u) => u.token === token);
      if (!user) throw new Error('Not authorized');
      return user;
    },
  },
};
```

Inside our `login` function, we search for a user in the database using simple password and username mechanisms. If we find a user, we can return its `token` value.

On the other hand, inside the `register` resolver, we check whether the user exists and, if not, we create one with a generated `_id` and `token`.

The `Mutation.user` resolver is the one used to pass the user object to `AuthorizedUserMutation`. It checks whether the token is provided via headers inside a request.

We also have a `Query.user` resolver that provides the same functionality as the `Mutation.user` resolver and the `AuthorizedUserMutation.changePassword` resolver.

> **Note**
> You can check the full resolver code in the GitHub repository.

Supergraph

At the root level of the repository, there is a `schema.graphql` file that is generated out of `src/todos/schema.graphql` and `src/users/schema.graphql`. This is the supergraph. Let's take a look at its generated code:

```
type Todo{
  _id: String!
  content: String!
  done: Boolean
}

type TodoOps{
  markDone: Boolean
}

type User{
  _id: String!
  username: String!
}

type AuthorizedUserMutation{
  createTodo(content: String!): String!
  todoOps(_id: String!): TodoOps!
  changePassword(newPassword: String!): Boolean
}

type AuthorizedUserQuery{
  todos: [Todo!]
  todo(_id: String!): Todo!
  me: User!
}

type Mutation{
  user: AuthorizedUserMutation!
  login(username: String!password: String!): String!
  register(username: String!password: String!): String!
}

type Query{
  user: AuthorizedUserQuery!
}
```

Inside this schema, you can see two schemas merged using all the federation rules.

Before we run the backend, we can see the code of `src/resolvers.ts` to check how the resolvers of our schemas merged. Here is the code:

```
import { mergeAxolotls } from '@aexol/axolotl-core';
import todosResolvers from '@/src/todos/resolvers.js';
import usersResolvers from '@/src/users/resolvers.js';

export default mergeAxolotls(
  todosResolvers,
  usersResolvers
);
```

Here, we imported the resolver files from each service and merged them using `mergeAxolotls` functions.

There is also an `axolotl.json` file that is responsible for the federation specification. Let's take a look at its code:

```
{
    "schema": "schema.graphql",
    "models": "src/models.ts",
    "federation":[
        {
            "schema":"src/todos/schema.graphql",
            "models":"src/todos/models.ts"
        },
        {
            "schema":"src/users/schema.graphql",
            "models":"src/users/models.ts"
        }
    ]
}
```

Here, we specified the main schema and models. Then, in the `federation` field, we specified all the schemas that make up the final schema.

Running federated service

Now that we have examined all the code responsible for our federated service, let's run it and test it.

To do so, execute the following command:

```
npm run dev
```

This will start the Yoga server on port 4002 with some example queries prepared to test the service.

Enter `http://localhost:4002/graphql` and play with GraphiQL. You should see something like this:

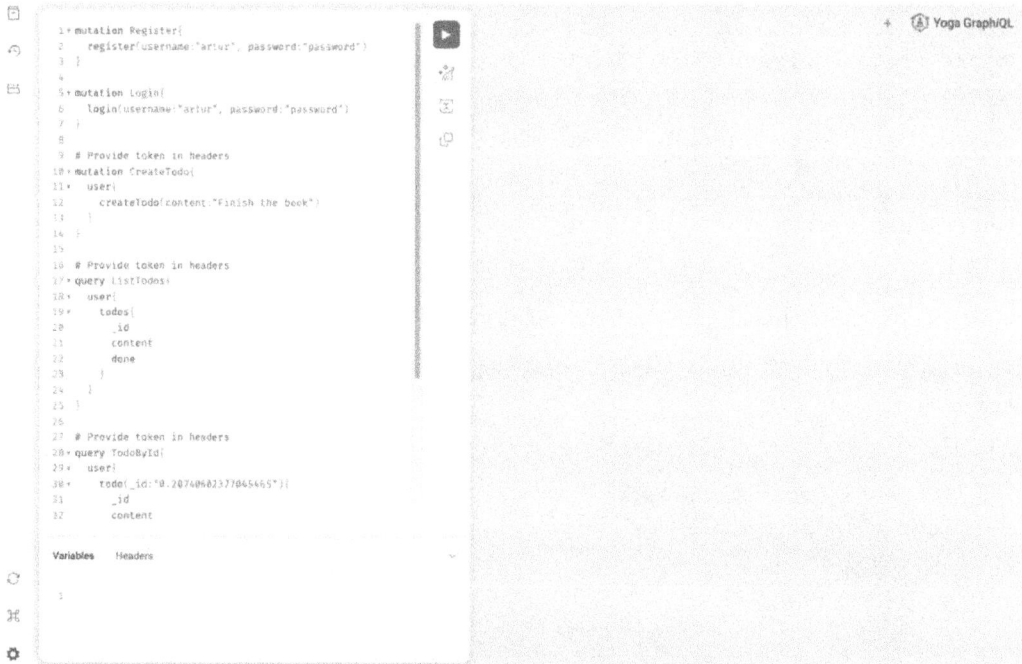

Figure 7.3: GraphiQL with our supergraph to-do list

You can try executing all the queries and mutations in the GraphiQL interface now.

Remember that to access the `Authorized` queries and mutations, you need to provide the token returned from the `login` or `register` function inside the GraphiQL **Headers** section, like so:

```
{
    "token":"YOUR_TOKEN_HERE"
}
```

With that, we have implemented and executed a GraphQL Federation-based service with two schemas being joined into one supergraph. In the next section, we will discuss the current and future state of GraphQL Federation.

Meeting the need for GraphQL Federation

The OpenFederation Specification is an open specification that aims to make federation in GraphQL a standardized feature, instead of being confined to any proprietary solution such as Apollo Federation. The goal is to standardize how GraphQL schemas declare their federation capabilities and establish a universal way to build gateway implementations.

The Open Federation Specification, which is still in draft status at the time of writing, is designed to evolve GraphQL Federation in a way that it can be adopted across all GraphQL runtimes and tools, ensuring greater interoperability and consistency.

Here are some principal aspects defined in the specification:

- **Federated schema directives**: Details how to use certain directives such as `@key`, `@requires`, `@provides`, and `@extends` to establish relationships across types and services.
- **Type system extensions**: Defines a set of type system extensions that need to be added by all service providers to implement Open Federation. These include `_FieldSet` and `_Entity` types.
- **Query planning**: Establishes how to construct an execution plan for a federated query, which includes splitting a client query into separate operations for each service and incorporating data from one service into another.
- **Schema validation**: Provides a set of conformance demands on federated schemas to ensure they abide by the principles of GraphQL Federation.
- **Federation capabilities**: It calls out which federation features each service supports. It gives a direction to services to declare their exact federation capabilities instead of indirectly implying them.

In general, the Open Federation Specification is an important advancement aiming to make GraphQL Federation more universally adoptable for building more efficient, robust GraphQl APIs. The open source and collaborative nature of the specification ensures it caters to the diverse needs of blog developers and promotes best practices across the GraphQL community.

Summary

In this chapter, we learned what GraphQL Federation is, as well as its key concepts. In essence, GraphQL Federation empowers teams to work independently and yet together, enabling more rapid development and scaling. This makes it especially vital in large-scale, enterprise-grade projects, making it a vital part of the GraphQL ecosystem.

We also saw how to merge schemas together, along with some examples.

Currently, the execution of federation in your project is open to your discretion. It is worth noting that federation is a relatively new topic that requires further development. For instance, there may be differing opinions on certain federation concepts, such as resolving arguments using intersection. Instead, one alternative approach could be to route a separate query with the exact argument to a dedicated schema. Ultimately, there is a need for ongoing discussions and efforts to establish a unified standard for federation in the future.

In the next chapter, we will see how to execute schema-first systems.

Part 4 - Advanced GraphQL

In this part, you will learn how to create a small backend and frontend for the Questions and Answers system. Then we will dive deeper into GraphQL security, error handling, and documentation. In the end, I will teach you how to use visualization to read and improve your schemas.

This part contains the following chapters:

8

Executing Schema-First Systems

We have already learned a lot throughout this book, so it's time to do something that will work both on the backend and frontend. In this chapter, we will learn how to design a schema and implement it using TypeScript and a Node.js backend for a Questions and Answers system. This app will be built on a GraphQL schema, work with a MongoDB database, and include a username-password authentication system.

Just using GraphQL Schema Definition Language to describe a system isn't very useful. To make full use of GraphQL and this schema-driven software, you need to code a GraphQL server. To do this, we need to provide resolver functions for GraphQL-type fields, which can be done by writing the backend server for the GraphQL schema.

In this chapter, we will cover the following topics:

- Setting up our project
- Creating the schema
- Connecting a database schema and a type system
- Handling user authorization and authentication
- Adding question and answer types
- Bringing the resolvers together

Technical requirements

To complete this chapter, you will need the following:

- Terminal and Node.js installed

- Docker installed (`https://docs.docker.com/engine/install/`)

- A GraphQL IDE such as VS Code, Sublime Text, Notepad++, or GraphQL Editor

- A basic knowledge of TypeScript

You can access the code in this chapter on the book's GitHub repository: `https://github.com/PacktPublishing/GraphQL-Best-Practices/tree/main/chapter-08`.

Setting up our project

As mentioned, we will create a forum service for a Questions and Answers system. Our system will allow users to do the following:

- Register a user account

- Log in and out

- Post a question on a public board

- Reply to a question

- Reply to an answer

- Vote for a question

To code our project, we will use the same backend stack we used in previous chapters. So, we will start by bootstrapping the project. Open your terminal and enter the following command:

```
npx @aexol/axolotl create-yoga
```

Then ,enter the project folder and install the following additional dependencies:

```
npm i mongodb i-graphql jsonwebtoken
```

Let's quickly review the dependencies here:

- **MongoDB** is a popular NoSQL database that provides a flexible and scalable solution to store and retrieve data. It uses a document-oriented model, where data is stored in flexible and schema-less documents, allowing for easy and dynamic data modeling.

- **Axolotl** is a code generation tool for backends that automatically creates TypeScript types based on your GraphQL schema. It eliminates the need for manual type definitions, saving you time and reducing the chances of errors. It works with GraphQL Yoga and Apollo Server.

- **JWT** (**JSON Web Token**) is a compact and self-contained tool used to securely transmit information between parties. It consists of three parts – a header, a payload, and a signature. The header contains metadata, the payload contains user claims, and the signature ensures integrity.

So, we need `mongodb` to connect to our database and store some information there, we need `i-graphql` used together with `axolotl` to ensure type safety, and we need `Jsonwebtoken` to secure access to our backend.

Finally, we need to add the `mongo-dev` command to run Docker with MongoDB locally:

```
"mongo-dev": " docker run -p 27017:27017 --name=qa mongo:latest"
```

With everything in place, we can now move on to creating the schema.

Creating the schema

In this section, we will begin by creating a `schema.graphql` file in the project root, where we will define the structure of our system. Our first step will be to focus on user registration and login. Once users are registered, they will be able to contribute to our system by adding questions and answers. To ensure security, we will implement a simple username-password authentication system, storing the authentication data in our database.

Additionally, we will utilize the pipes mechanism to assign ownership of questions and answers to individual users. This approach will provide a seamless experience for both users and administrators.

Authentication and authorization

Before we design our login and register resolvers, we will create interfaces and a `User` type, responsible for the ownership of questions and answers. It should look like this:

```
interface StringId{
  _id: String!
}

interface Dated{
  createdAt: String!
  updatedAt: String!
}

interface Owned{
  user: User!
}
type User implements StringId & Dated{
  username: String!
```

```
  _id: String!
  createdAt: String!
  updatedAt: String!
}
```

There are three interfaces:

- The `StringId` interface is responsible for holding the database ID. Every type implementing this interface will have an `_id` field of type `String!`.

- The `Dated` interface is responsible for adding `createdAt` and `updatedAt` fields, so we know when the object was created or updated.

- The `Owned` interface should be implemented by every type that needs `User` to be created. As previously mentioned, every question and answer is owned by a `User` object, so we will need an interface to make it clear which questions the user has asked and which answers they have posted.

Then, the `User` type holds the user's username. Inside our database, it will also hold a password that has been encrypted using a hash and salt mechanism. We should not expose these fields in GraphQL to improve security in our application (if they were exposed, it would create the risk of allowing others to see everybody's password when returning `Question` and `Answer` objects owned by a `User` object).

With the interfaces and `User` type in place, we can move on to the authorization part and write queries and mutations for them:

```
type AuthPayload{
  token: String!
  user: User!
}

type PublicMutation{
  register(
    username: String!
    password: String!
  ): AuthPayload!
  login(
    username: String!
    password: String!
  ): AuthPayload!
}

type Query{
  me: User!
}
```

```
type Mutation{
  public: PublicMutation
}
```

Inside our `PublicMutation` type, we create `login` and `register` fields to handle our username-password authentication. They both require the same input and return `AuthPayload`, which consists of the token needed for the authorization of owned pipes later on. We named this mutation `PublicMutation` because it doesn't require any authorization.

In order to enhance the functionality of our system, we have included a me field in our `Query` type. This field serves the purpose of returning the user's data when they are logged in and authorized with a valid token. By utilizing this feature, we ensure that only authenticated users are granted access to their personal information, resulting in a more secure and personalized experience.

With the authentication and authorization part done, we can move on and design the schema for the actual Questions and Answers system.

Questions and answers

Each question and answer in our system represents a message from the user. This means that every interaction and contribution made by the user, whether it is asking a question or providing an answer, is treated as a message within our system.

To handle that inside the schema, we need a dedicated interface that will hold the shared fields between `Question` and `Answer`. We will create the `Message` interface to control these properties:

```
interface Message implements StringId & Dated & Owned{
  content: String!
  score: Int!
  _id: String!
  createdAt: String!
  updatedAt: String!
  user: User!
  answers: [Answer!]!
}
```

So, every question and answer should have text content provided by the user, as well as a score to sort the best questions and answers to the top of the list. We also add common interfaces such as `StringId` and `Dated` to our `Message` interface. The `Message` interface implements the `Owned` interface to communicate that every `Message` is created and owned by a `User` object.

Now, we can write the actual GraphQL code implementing the `Message` interface. The following are the `Question` and `Answer` types:

```
union ToAnswer = Question | Answer
type Question implements Message & StringId & Dated &
Owned{
  content: String!
  score: Int!
  _id: String!
  answers: [Answer!]!
  title: String!
  createdAt: String!
  updatedAt: String!
  user: User!
}

type Answer implements Message & StringId & Dated & Owned{
  content: String!
  score: Int!
  _id: String!
  to: ToAnswer
  createdAt: String!
  updatedAt: String!
  user: User!
  answers: [Answer!]!
}
```

As you can see, the two types implement four interfaces. I highlighted the fields that make the difference between the `Question` and `Answer` types, which are as follows:

- The `Question` type has the `title` field, which holds the shortened version of the question

- The `Answer` type has the `to` field, which is responsible for the relationship between the answer and the question, or the answer and another answer

In order to facilitate the creation of questions and answers, we will implement operations dedicated to these tasks. To ensure proper authorization and ownership, we will group these mutations and place them behind ownership pipes. This approach will allow us to maintain control over the content creation process, ensuring that only authorized users can create and manage questions and answers within our system:

```
input CreateQuestion{
  content: String!
  title: String!
}
```

```
input CreateAnswer{
  content: String!
  title: String!
  to: String!
}

type UserMutation{
  postQuestion(
    createQuestion: CreateQuestion!
  ): String
  postAnswer(
    createAnswer: CreateAnswer!
  ): String
   vote(
    _id: String!
  ): Int
}
type Mutation{
  user: UserMutation
  public: PublicMutation
}
```

Let's break this code down:

- First, we have the UserMutation type – its fields can only be resolved if Mutation.user passes the authorization phase.

- Then, we have the UserMutation.postQuestion resolver, which is used to create a question by an authorized user using the CreateQuestion input.

- We also have the UserMutation.postAnswer resolver, which is used to answer a question or reply to another answer using the CreateAnswer input.

- The last resolver, UserMutation.vote, is used to increment the score of the answer or question.

The only part left is the part that doesn't need user authorization and is available without a user logging in – *posting* questions and answers should require the user to have an account; however, just *viewing* the discussions shouldn't require one. We want users who don't have an account to still see the question by its _id or be able to search for questions. So, here is the part of the schema responsible for the public query:

```
type Query{
  search(
    query: String!): [QuestionsResponse!]!
```

```
    top: [QuestionsResponse!]!
  question(
    _id: String!): Question
    me: User!
}

type QuestionsResponse{
  question: Question!
  bestAnswer: Answer
}
```

We have three new fields on the `Query` type. The `search` field should return all the questions that match the query parameter. We return it as an array of the `QuestionResponse` type. In order to provide comprehensive search results, we decided to include both the question and the best answer in the displayed results. By including both pieces of information, users will be able to easily access relevant content and have a more complete understanding of the search results.

We also added the `top` field to display top questions that should be listed on the home page, as well as a `question` field, which queries questions by their `_id` to return a matching question object.

Overall, to build this schema, we used both ownership and logical pipes, as well as a lot of interfaces to inform consumers and schema developers which types share which fields. The following diagram shows the schema graph:

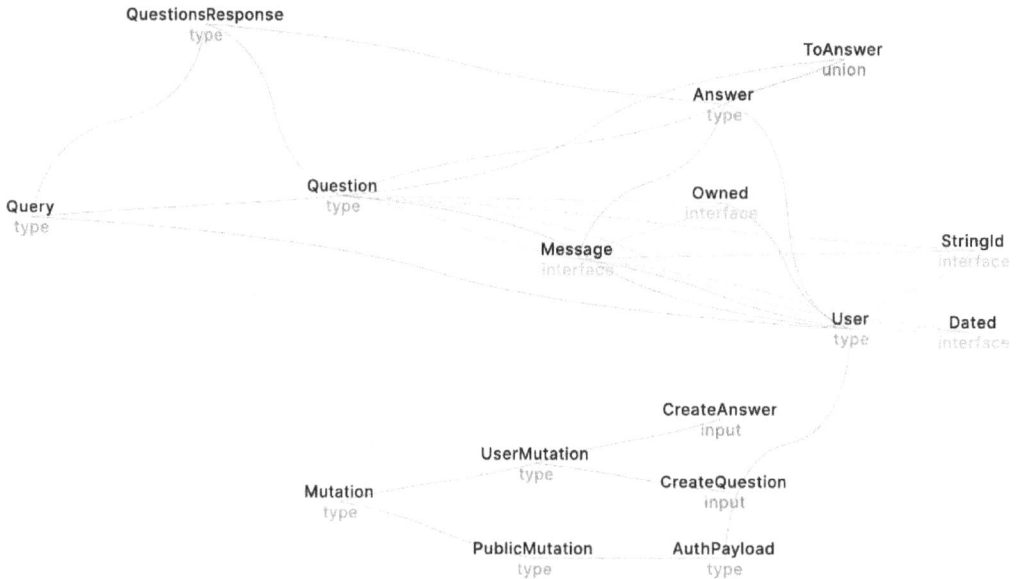

Figure 8.1: The schema graph diagram

This schema should be ready for schema-first backend development. In the next section, we will connect our schema to the database and type system.

Connecting the schema, database, and type system

Before we start developing the resolvers, it would be good to have a database and TypeScript types in place. We will start by executing the `models` command (which we added in the *Setting up our project* section) to `package.json`. This command generates TypeScript types for our GraphQL Schema to utilize schema-first development:

```
npm run models
```

This will generate a TypeScript type system, based on our GraphQL schema inside the `src/models.ts` file.

After that, we will create a file to specify the Mongo models. Let's put it inside the `src/orm.ts` file:

```typescript
import { Answer, Question, User } from "@/src/models.js";
import { iGraphQL, MongoModel } from "i-graphql";
import { ObjectId } from "mongodb";

export type UserModel =  MongoModel<User> & {
  passwordHash:string;
  salt:string;
}
export type QuestionModel = MongoModel<Question>
export type AnswerModel = MongoModel<Answer>

export const orm = async () => {
  return iGraphQL<
    {
      Question: QuestionModel,
      Answer: AnswerModel,
      User: UserModel
    },
    {
      _id: () => string;
      createdAt: () => string;
      updatedAt: () => string;
    }
  >({
    autoFields:{
      _id: () => new ObjectId().toHexString(),
      createdAt: () => new Date().toISOString(),
```

```
        updatedAt: () => new Date().toISOString(),
    }
  });
};

export const MongOrb = await orm();

MongOrb("Question").collection.createIndex({
  "content":"text",
  "title":"text"
})
```

First of all, we created a type-safe **object-relational mapping (ORM)** system with the i-graphql library and provided it with our axolotl models. Inside the generated Zeus files, there are types that can help us construct model definitions. Specifically, ModelTypes is the type containing all the GraphQL types, interfaces, unions, scalars, enums, and inputs with TypeScript types.

> **Note**
>
> ORM is a technique that allows developers to seamlessly connect and interact with databases. It acts as a bridge between a database and application code, abstracting away the complexity of database operations.

Then, for Question and Answer types, we just used the built-in MongoModel function to convert fields that are not scalar into string types.

After that, we added automatic fields that can be autogenerated during operations – this included _id for the generation of a unique identifier for each object, createdAt to store the date when an object was created, and updatedAt to store the last update date.

Then, we exported the MongOrb variable so that we could use our GraphQL-based orm system across the project.

Finally, we added a search index to the content and title fields to allow the searching of the database, using those two fields with the Question type.

The last thing we need to do to complete our connection is to generate the axolotl models that are used inside the resolvers. Run the following command:

```
npm run models
```

This command uses axolotl to generate models used for createResolvers type safety.

Now that we have successfully set up our TypeScript types, database, and ORM, we can proudly move on to writing resolvers. This accomplishment marks a significant milestone in the development process and allows us to begin implementing the logic and functionality that will drive our system.

Handling user authorization and authentication

We have now prepared our schema to use standard username-password authentication together with JWT authorization, using pipes. We will now implement authentication resolvers and authorization mechanisms.

To begin, create a file called `src/auth.ts` – this is where we will keep all the functions connected to authentication and authorization.

Then, we will create some helper functions to keep our authorization mechanisms secure. To do so, we need to store the hashed password inside the database:

```
import crypto from 'node:crypto';
import jwt from "jsonwebtoken";
import { createResolvers } from '@/src/axolotl.js';
import { MongOrb } from '@/src/orm.js';
import { GraphQLError } from 'graphql';


const secretKey = 'your-secret-key';

const passwordSha512 = (password: string, salt: string) => {
  const hash = crypto.createHmac('sha512', salt);
  hash.update(password);
  const passwordHash = hash.digest('hex');
  return {
    salt,
    passwordHash,
  };
};
```

This function is used to generate `passwordHash` using Node.js's built-in `crypto` library, and then `salt` is provided as the function parameter.

The password hash salt mechanism is a technique used to enhance the security of passwords stored in a database. It involves adding a random and unique piece of data, called a *salt*, to each password before hashing it. This ensures that even if two users have the same password, their hashed values will be different, making it harder for attackers to crack passwords through techniques such as rainbow tables or precomputed hashes.

The `salt` value is typically stored alongside the hashed password, allowing for the verification of passwords during the login process, like so:

```
const comparePasswords = ({ password, hash, salt }: {
  password: string;
  hash: string;
  salt: string
}) => {
```

```
    return hash ===
      passwordSha512(password, salt)
        .passwordHash;
  };
```

Additionally, we need to code the `comparePasswords` function to compare the password provided by the user during the login process with the one stored in the database.

As we have prepared all the functions, we can move on and implement the authentication and authorization mechanisms.

Authentication resolvers

To handle authentication, we have dedicated resolvers within the `PublicMutation` type of our schema. These resolvers will be housed in the `src/PublicMutation.ts` file. By centralizing these authentication-related resolvers in a separate file, we can easily manage and maintain the code specific to this functionality. Let's take a closer look at the code block in which these resolvers are implemented:

```
export const PublicMutation = createResolvers({
  PublicMutation: {
    register: async (_, { username, password }) => {
      const userExists = await MongOrb("User")
        .collection
        .findOne({username})
      if (userExists)
        throw new GraphQLError(
          "User already exist with this username"
        )
      const s = crypto.randomBytes(8).toString('hex');
      const {passwordHash, salt} =
        passwordSha512(password, s)
      const user = await MongOrb("User")
        .createWithAutoFields(
          "_id",
          'createdAt',
          'updatedAt'
        )({
          username,
          passwordHash,
          salt
        })
      const token = jwt.sign(
        { userId:user.insertedId },
        secretKey
```

```
      );
      return {
        token,
        user,
      };
    },
    login: async (_, { username, password }) => {
      const user = await MongOrb("User")
        .collection
        .findOne({
          username
        })
      if (!user) {
        throw new GraphQLError('User not found');
      }
      const validPass = comparePasswords({
        password,
        hash:user.passwordHash,
        salt:user.salt
      })
      if(!validPass){
        throw new GraphQLError('Invalid password')
      }
      const token = jwt.sign(
        { userId: user._id },
        secretKey
      );
      return {
        token,
        user,
      };
    },
  },
})
```

If we examine the resolver code more closely, we can see that our focus is primarily on the registration and login functionality. These resolvers handle the process of registering new users and authenticating existing users within our system.

During the `register` resolver, when the user creates a username, we check whether the user already exists in the database. If they exist, we throw an error:

```
const userExists = await MongOrb("User")
  .collection
```

```
  .findOne({username})

if(userExists)
  throw new GraphQLError(
    "User already exist with this username"
  )
```

Next, we generate `salt`, using the `crypto` library, and `passwordHash`, using our functions:

```
const s = crypto.randomBytes(8).toString('hex');
    const {passwordHash,salt} =
      passwordSha512(password,s)
```

After that, we specify that we want our `orm` system to generate the `_id`, `createdAt`, `updatedAt` fields, and we provide the `username`, `passwordHash`, and `salt` values to store this user inside the database:

```
const user = await MongOrb("User")
  .createWithAutoFields(
    "_id",
    'createdAt',
    'updatedAt'
  )({
    username,
    passwordHash,
    salt
  })
```

Then, we create a signed token the using `jsonwebtoken` library and return it, just as we would return the `login` resolver with a token and a user instance:

```
const token = jwt.sign(
  { userId:user.insertedId },
  secretKey
);
return {
  token,
  user,
};
```

This token will later be used by the schema consumer to authorize their requests when operating on the `Question` and `Answer` types.

Moving on from the `register` resolver, let's look at the `login` resolver, which is shorter:

```
const user = await MongOrb("User")
  .collection
  .findOne({username})
if (!user) {
  throw new GraphQLError('User not found');
}
```

We start by checking whether the user exists inside the database. If they don't exist, we throw an error.

Then, we need to check whether the password is correct, which is done with the `comparePasswords` function:

```
const validPass = comparePasswords({
  password,
  hash:user.passwordHash,
  salt:user.salt
})
if(!validPass){
  throw new GraphQLError('Invalid password')
}
```

The last part of the login process is to return the token and user object, like in the `register` resolver:

```
const token = jwt.sign({ userId: user._id }, secretKey);
return {
  token,
  user,
};
```

Now, we have authentication in place, but we need to remember that it is behind the `Mutation.public` logical pipe; we will implement it after we finish implementing the authorization mechanism.

The authorization mechanism

During the JWT authorization process, the inclusion of the JWT token in the `Bearer` scheme of the `Authorization` header is a crucial step. This involves attaching the JWT to the header (using the `Bearer` prefix) to authorize the user. The server receives the JWT from the header, verifies its signature, and extracts the necessary information to authenticate and authorize the user, granting access to protected resources based on the claims within the JWT.

For this, we need some functions. First, we need a function to decode the signed token and retrieve some data from it:

```
export const decodeToken = (token: string) => {
  const verifiedToken = jwt.verify(token, secretKey);
  if (typeof verifiedToken !== 'object') {
    throw new GraphQLError('Token is not an object');
  }
  if (!verifiedToken.userId) {
    throw new GraphQLError('Invalid token');
  }
  return verifiedToken as { userId: string };
};
```

Here, we use the jsonwebtoken library to decode the token and then check whether it holds the userId property needed to validate the user.

Next, we need some helper functions to check whether userId inside the token exists inside the database, throwing an error if it doesn't:

```
export const getUser = async (
  authorizationHeader: string
) => {
  const { userId } = decodeToken(authorizationHeader);
  const user = await MongOrb("User").collection.findOne({
    _id: userId,
  });
  return user;
};

export const getUserOrThrow = async (
  authorizationHeader:string
) => {
  const user = await getUser(authorizationHeader);
  if (!user) {
    throw new GraphQLError('You are not logged in');
  }
  return user;
};
```

In the first helper function, getUser, we decode userId from our authorization header and return the user if it exists. The second function, getUserOrThrow, is used to throw an error when the user does not exist.

Now, we have prepared the functions to authorize a user. Next, we will write the resolvers responsible for checking whether the user is authorized.

Mutation pipes

In this section, we will implement the logical pipe for `Mutation.public`, which serves the purpose of returning an empty object. Additionally, we will incorporate an authorization-required pipe resolver specifically for the `Mutation.user` field, which will ensure that the user is authorized before entering their data. We will put those in the `src/Mutation.ts` file, like so:

```
import { getUserOrThrow } from '@/src/auth.js';
import { createResolvers } from '@/src/axolotl.js';
import { GraphQLError } from 'graphql';

export const Mutation = createResolvers({
  Mutation: {
    public:() => {
      return {}
    },
    user:async ([,,context]) => {
      const authHeader = context.request.headers
        .get("Authorization")
      if(!authHeader)
        throw new GraphQLError(
          "You must be logged in to use this resolver"
        )
      return getUserOrThrow(authHeader)
    }
  },
})
```

Our public mutation resolver for the `Mutation.public` field is just an empty logical pipe, simply returning an empty object to allow the further resolution of GraphQL. If we didn't return anything, the resolution would just stop. This is important – even if your logical pipe does nothing but group resolvers, it still has to return something!

Conversely, the `Mutation.user` resolver extracts the header from the request and passes it to the `getUserOrThrow` function. Then, the function returns a user or throws an error if the user is not authorized or the user token is invalid. This is also important – we need to return the user in this resolver, as it will be used by all `UserMutation` fields later.

And that's it! With these GraphQL mutations and resolvers, you can now create a username, password, and JWT-based login and user registration system for your application. Make sure to incorporate authentication middleware to protect your sensitive endpoints and validate the JWT token before granting access to protected resources.

Now, let's move on to coding the resolvers to post questions and answers.

Adding question and answer types

In this section, I will show you how to use ownership pipes to check whether a user is authorized and attach the information about the owner to the actual object. We will learn how to code basic database operations for our GraphQL fields and explore how to deal with relations between different types.

Coding user mutations

We will begin by writing mutations that allow us to insert data into a database. Create the `src/UserMutation.ts` file, import the required functions, and then create a `postQuestion` resolver:

```
import { MongOrb, UserModel } from '@/src/orm.js';
import { createResolvers } from '@/src/axolotl.js';
import { GraphQLError } from 'graphql';

export const UserMutation = createResolvers({
  UserMutation:{
    postQuestion: async ([source],args) => {
      const userSource = source as UserModel
      const result = await MongOrb("Question")
        .createWithAutoFields(
          '_id',
          'createdAt',
          'updatedAt'
        )({
          ...args.createQuestion,
          score: 0,
          user:userSource._id,
          answers:[],
        })
      return result.insertedId
    },
  }
});
```

Currently, our backend uses GraphQL Yoga; however, this tool masks unexpected errors by default to keep the system secure. Furthermore, here, we also add `GraphQLError`, which makes errors visible to the schema consumers. This is because some errors *should* be visible – for example, when you register as a user and your username is taken (we will dive into this subject in more detail in *Chapter 11*).

Let's move on to the `postQuestion` resolver implementation; as you can see, we extract the source argument. This is because we need to know the ID of the user who will be the owner of `Question`. Then, within the function body, we cast the source coming from the previous resolver to `UserModel`.

Next, we use our `orm` system's `createWithAutoFields` function to create the `Question` object, together with auto-creatable arguments such as `_id`, `createdAt`, and `updatedAt`. Then, we provide `args.createQuestion` and use spread notation to insert all fields from the input directly into the object.

When the question is created, we set the `score` to 0, as this is a newly created question. It also has no answers, so we provide an empty array. Since the resolver is expected to return a string value, we opt to return `insertedId`, which is of type `String` in our case. By returning this value, we confirm the successful insertion operation and allow for further processing or feedback, based on `insertedId`.

We also need the `postAnswer` resolver to allow a user to reply to other questions and answers:

```
postAnswer: async ([source], args) => {
  const userSource = source as UserModel;
  const [q, a] = await Promise.all([
    MongOrb('Question').collection.findOne({
      _id: args.createAnswer.to,
    }),
    MongOrb('Answer').collection.findOne({
      _id: args.createAnswer.to,
    }),
  ]);
  const object = q || a;
  if (!object) {
    throw new GraphQLError(
      `Question or Answer with id
        "${args.createAnswer.to}" does not exist
          anymore.`
    );
  }
  const result = await MongOrb('Answer')
    .createWithAutoFields(
      '_id',
      'createdAt',
      'updatedAt',
    )({
```

```
        answers: [],
        content: args.createAnswer.content,
        score: 0,
        user: userSource.username,
        to: object._id,
      });
    return result.insertedId;
  },
```

While the postAnswer resolver looks similar to the postQuestion resolver, the difference is we need to find the question or answer by its _id before creating an Answer object. To do so, we use the Promise.all function to fire two promises at the same time, only using the result of the one that returns a value. If nothing is found, we throw an error.

Lastly, we need a UserMutation resolver that will allow users to vote (as many times as they want) on a question or answer:

```
  vote: async ([source], args) => {
    const q = await MongOrb('Question')
      .collection
      .findOne({
        _id: args._id,
      });
    if (!q) {
      const a = await MongOrb('Answer')
        .collection
        .findOne({
          _id: args._id,
        });
      if (!a)
        throw new GraphQLError(
          `Question or Answer with id "${args._id}" does
            not exist anymore.`
        );
      await MongOrb('Answer').collection.updateOne(
        { _id: args._id },
        {
          $inc: {
            score: 1,
          },
        },
      );
      return a?.score + 1;
    } else {
```

```
    await MongOrb('Question').collection.updateOne(
      { _id: args._id },
      {
        $inc: {
          score: 1,
        },
      },
    );
    return q?.score + 1;
  }
},
```

In this resolver, we also find a `Question` or `Answer` type with the `_id` parameter provided by the schema consumer. Then, we update the object in the database using MongoDB's `$inc` parameter – we use the `$inc` parameter to avoid a race condition and ensure that every vote will increase the overall score.

> **Note**
>
> A **race condition**, also known as a race hazard, refers to a situation within a system where the system's behavior is influenced by the specific sequence or timing of uncontrollable events. This can result in unexpected or inconsistent outcomes that are not in line with the intended behavior of the system.

Now, we can create objects within our system. Next, we need to provide a way to query them.

Writing queries

As we are able to register users and post questions and answers, we need to be able to retrieve them from our GraphQL backend. To do this, we will write four queries that will be available publicly without authorization, placing them in a separate file called `src/Query.ts` where the root resolvers of `Query` will also be placed:

```
import { MongOrb } from '@/src/orm.js';
import { createResolvers } from '@/src/axolotl.js';
import { getUserOrThrow } from '@/src/auth';
import { GraphQLError } from 'graphql';

export const Query = createResolvers({
  Query: {
    question:(_,args) => {
      return MongOrb('Question')
        .collection
```

```
        .findOne({_id:args._id})
    },
    search: async (_,args) => {
      const questions = await MongOrb("Question")
        .collection
        .find(
          {$text:{$caseSensitive:false,
            $search:args.query}},
          {sort:{score:1}} // sorting by score
        )
        .toArray()
      return questions.map(question => ({
        question
      }))
    },
    me:([_,__,context]) => {
        const authHeader = context.request.headers
          .get("Authorization")
        if(!authHeader)
          throw new GraphQLError(
            "You must be logged in to use this resolver"
          )
        return getUserOrThrow(authHeader)
    },
    top:() => {
      return MongOrb('Question')
        .collection
        .find(
          {},{
            sort:{score: 1},limit: 10}
        )
        .toArray()
    }
  },
});
```

Here are the four root queries:

- The first resolver, Query.question, takes the _id argument and tries to find the question by its _id field in the database.

- The second resolver, Query.search, uses the MongoDB search index to search the collection with the given query. Then, it returns the best results. It also returns the answer with the most votes, so we can display it later in search results together with the question.

- The third resolver, `Query.me`, has the same code as `Mutation.user`, but this is used to return the user and resolve it, so the schema consumer can display the user's username somewhere in the UI.

- The last resolver, `Query.top`, returns the top 10 questions with the biggest score.

We have implemented all the queries and mutations, but not all the resolvers.

Relation resolvers

It looks like we've just finished implementing the backend part of the project, but something is missing. Let's take a look at the `Question` type again:

```
type Question implements Message & StringId & Dated &
Owned{
  content: String!
  score: Int!
  _id: String!
  answers: [Answer!]!
  title: String!
  createdAt: String!
  updatedAt: String!
  user: User!
}
```

The `Answer` and `User` fields are not held as objects but as strings, containing the IDs of objects inside the database. So, we need resolvers for those fields to make it possible to execute this query:

```
query Search($question__id: String!) {
  question(_id: $question__id) {
    content
    score
    answers {
      content
    }
    title
    user {
      username
    }
  }
}
```

In a typical scenario, without using GraphQL resolvers on the `Question.user` field, an error would occur because the system would not fetch the `username` property from the user's `_id` directly. Instead, it requires the entire `User` object to be retrieved.

The same principle applies to the `Question.answers` field as well. By using GraphQL, we can ensure that the complete `User` object is fetched, including the necessary properties such as username, to avoid any potential errors and ensure seamless data retrieval and manipulation.

To allow us to fetch the related objects, we will write resolvers for the `user` and `answers` fields. Create the `src/Question.ts` file and add the following content inside:

```
import { MongOrb, QuestionModel } from '@/src/orm.js';
import { createResolvers } from '@/src/axolotl.js';

export const Question = createResolvers({
  Question:{
    user: async ([source]) => {
      const questionSource = source as QuestionModel
      return MongOrb("User").collection.findOne({
        username: questionSource.user
      })
    },
    answers: async ([source]) => {
      const questionSource = source as QuestionModel
      return MongOrb("Answer")
        .collection
        .find(
          {to: questionSource._id}
        )
        .toArray()
    }
  },
});
```

The `Question.user` resolver returns the user who created the question, gathering the information from the source (the source is the data returned from the previous resolver that returned the `Question` type). Then, the `Question.answers` resolver gives us all the answers from the database based on the question `_id`.

We also need those resolvers for the `Answer` type, as this also has some relation-based fields inside its type definition:

```
type Answer implements Message & StringId & Dated & Owned{
  content: String!
  score: Int!
  _id: String!
  to: ToAnswer
  createdAt: String!
  updatedAt: String!
```

```
    user: User!
    answers: [Answer!]!
}
```

The Answer type holds the same relational fields (User and Answer) inside its type definition, but it also has the Answer.to field, which is a union of Question and Answer and can return a question or answer object.

After creating the src/Question.ts file, we will now create a src/Answer.ts file and add the following code inside it:

```
import { AnswerModel, MongOrb } from '@/src/orm.js';
import { createResolvers } from '@/src/axolotl.js';

export const Answer = createResolvers({
  Answer: {
    user: async ([source]) => {
      const answerSource = source as AnswerModel;
      return MongOrb('User').collection.findOne({
        username: answerSource.user,
      });
    },
    answers: async ([source]) => {
      const answerSource = source as AnswerModel;
      return MongOrb('Answer')
        .collection.find({
          to: answerSource._id,
        })
        .toArray();
    },
    to: async ([source]) => {
      const answerSource = source as { to: string };
      const [q, a] = await Promise.all([
        MongOrb('Question').collection.findOne({
          _id: answerSource.to,
        }),
        MongOrb('Answer').collection.findOne({
          _id: answerSource.to,
        }),
      ]);
      return q || a;
    },
  },
});
```

The resolver functions for `Answer.user` and `Answer.answers` closely resemble the resolver functions for the `Question` type. However, within the `Answer.to` resolver, we need to examine both the `Question` and `Answer` collections to determine what the `Answer` object addresses. This is necessary because the `to` field is of a union type and requires us to consider both collections to determine whether a question or answer is returned.

Now that we have implemented all the relation resolvers needed inside this project, we can move on and get everything running as a server.

Bringing the resolvers together

Since we have implemented all the resolvers in separate files, it is essential to find a way to provide them to our GraphQL server. To do so, create a `src/index.ts` file with the following content:

```
import {
  graphqlYogaAdapter
} from '@aexol/axolotl-graphql-yoga';
import { createResolvers } from '@/src/axolotl.js';
import { UserMutation } from '@/src/UserMutation.js';
import { Query } from '@/src/Query.js';
import { Mutation } from '@/src/Mutation.js';
import { PublicMutation } from '@/src/PublicMutation';
import { Question } from '@/src/Question';
import { Answer } from '@/src/Answer';

const resolvers = createResolvers({
  ...Query,
  ...Mutation,
  ...PublicMutation,
  ...UserMutation,
  ...Question,
  ...Answer,
});

graphqlYogaAdapter({ resolvers })
  .server
  .listen(
  parseInt(process.env.PORT || '4000'),
  () => {
    console.log(
      'LISTENING to ' +
      (process.env.PORT || '4000')
    );
  });
```

Inside this file, we import all our resolvers from separate files and put them inside the `createResolvers` function. Then, we pass the object that has all the resolvers to `graphqlYogaAdapter` to run the server.

Now, to run the project, we must build it using the following command:

```
npm run build
```

Then, we can run the server:

```
npm run start
```

After that, you should be able to interact with GraphiQL on `localhost:4000/graphql`. To test that everything is working, follow these quick steps:

1. Register a user using the `publicMutation.register` query:

    ```
    mutation Register(
      $register_username: String!,
      $register_password: String!
    ) {
      public {
        register(
          username: $register_username,
          password: $register_password
        ) {
          __typename
          token
        }
      }
    }
    ```

2. Log in as that user and keep the response:

    ```
    mutation Login(
      $login_username: String!,
      $login_password: String!
    ) {
      public {
        login(
          username: $login_username,
          password: $login_password
        ) {
          token
        }
      }
    }
    ```

3. To ensure that the backend can identify the user making requests and grant them permission, we need to include the token received from the `Login` mutation in the `headers` field of the request. Specifically, we should add it as an `Authorization` header for all mutations within `UserMutation`. This way, the backend will have the necessary information to validate and authorize the requests made by the user:

```
mutation PostQuestion(
  $postQuestion_createQuestion: CreateQuestion!
) {
  user {
    postQuestion(
      createQuestion: $postQuestion_createQuestion
    )
  }
}

"variables": {
  "postQuestion_createQuestion": {
    "title": "Are we alone in space?",
    "content": "I wonder what happens when we find out
                another species"
  }
}

"headers":{
  "Authorization":"TOKEN_FROM_LOGIN"
}
```

> **Note**
> You can use `https://jwt.io/` to check whether your token is valid.

With the implementation of authentication and authorization mechanisms, we have established a reliable way to authenticate and authorize users as schema consumers. As a result, the backend system is now fully prepared to integrate with the frontend.

Summary

In this chapter, we created a Questions and Answers system, and through this, we learned how to create a fully working, database-connected, deployable backend. We started by creating the GraphQL schema for the Questions and Answers system, and then we implemented resolvers using GraphQL Yoga, Axolotl, iGraphQL, and MongoDB.

Implementing the full backend helped us to understand how the GraphQL preparation phase speeds up the implementation process.

In the next chapter, we will write a frontend that will communicate with the backend that we have created, allowing our users to interact with the application, and we will also learn the best practices to consume schema from the frontend.

9

Working on the Frontend with GraphQL

With the rise of GraphQL, developers have been able to build efficient and flexible APIs that cater to the specific needs of their applications. However, GraphQL is not just limited to the backend; it also offers incredible potential on the frontend.

One tool that greatly simplifies our interaction with GraphQL is **GraphQL Zeus**. This powerful code generator works straight out of the box, and helps us to automatically generate type-safe bindings for our GraphQL schema.

Using the Question and Answer system we created in the previous chapter, we will create a frontend to connect to the already established backend. We will also prepare a good communication layer between this frontend and backend.

By the end of this chapter, you will have a solid understanding of how to create a frontend application using GraphQL, GraphQL Zeus, Vite, and Shadcn.

So, in this chapter, we will cover the following topics:

- Setting up our project
- Preparing our GraphQL Layer
- Creating the system's frontend
- Creating authentication pages
- Crafting post question page
- Implementing questions page

> **Note**
> Psst, as this chapter is about the frontend, it will be a long one. Prepare yourself!

Technical requirements

To complete this chapter, you will need the following:

- Terminal and NodeJS installed

- A GraphQL IDE such as VSCode, Sublime Text, Notepad++, or GraphQL Editor

- A basic knowledge of TypeScipt and ReactJS

As mentioned in the introduction, we will be using the Question and Answer system from *Chapter 8*, so you should have completed the previous chapter too.

You can access the code in this chapter on the book's GitHub repository at `https://github.com/PacktPublishing/GraphQL-Best-Practices/tree/main/chapter-09`.

Setting up our project

Before we connect our frontend to the GraphQL backend that we created in the previous chapter, we need to bootstrap the frontend project to include some tooling and a design system.

Recently, Shadcn has become a really popular open source design system, which we will use for our project to make it look good enough without having to hire a designer. It uses ReactJS as a framework and is based on **TailwindCSS** for styling. If you are not familiar with these, here are the basic definitions:

- **ReactJS** is a JavaScript library that allows us to build **User Interfaces** (**UIs**) for web applications by creating reusable UI components. It follows a component-based architecture that promotes a declarative and efficient way of building interactive web applications.

- **TailwindCSS** is a utility-first CSS framework that provides a set of pre-designed utility classes. It enables developers to rapidly build responsive and customized UIs by composing these utility classes, without the need for writing custom CSS.

Then, to connect our frontend to the backend system, we will use GraphQL Zeus – this will generate a client that will connect to our backend without losing type-safety.

So, to set up all four elements, as well as the rest of the project, follow these steps:

1. Start by creating a project with Vite:

   ```
   npm create vite@latest
   ```

   When prompted, choose ReactJS and TypeScript within the interactive mode of this command.

2. Next, install `tailwind` and `postcss`:

   ```
   npm install -D tailwindcss postcss autoprefixer
   npx tailwindcss init -p
   ```

3. Add the following code to the `tsconfig.json` file to resolve the paths and avoid relative paths inside the project:

```
{
  "compilerOptions": {
    // ...
    "baseUrl": ".",
    "paths": {
      "@/*": [
        "./src/*"
      ]
    }
    // ...
  }
}
```

4. Install Vite's `tsconfig` paths plugin:

```
npm i -D vite-tsconfig-path
```

This plugin allows Vite to understand how to resolve paths that start with @ from the `tsconfig.json` file.

5. Set up the `vite.config.ts` file, including the `tsonfigPaths` and `react` plugins:

```
import { defineConfig } from 'vite'
import react from '@vitejs/plugin-react'
import tsconfigPaths from 'vite-tsconfig-paths';

// https://vitejs.dev/config/
export default defineConfig({
  plugins: [
    tsconfigPaths(),
    react(),
  ],
});
```

6. Initiate `shadcn`, and for each question, set the following options:

```
npx shadcn-ui@latest init
✔ Would you like to use TypeScript (recommended)? … yes
✔ Which style would you like to use? › Default
✔ Which color would you like to use as base color? › Stone
✔ Where is your global CSS file? … src/index.css
✔ Would you like to use CSS variables for colors? … no / yes
✔ Are you using a custom tailwind prefix eg. tw-? (Leave blank
if not) …
```

✔ Where is your tailwind.config.js located? … tailwind.config.js

✔ Configure the import alias for components: … @/components

✔ Configure the import alias for utils: … @/utils

✔ Are you using React Server Components? … **no** / yes

✔ Write configuration to components.json. Proceed? … yes

7. Next, install GraphQL Zeus:

```
npm i -D graphql-zeus
```

8. Add the zeus script to the package.json scripts:

```
"zeus": "zeus http://localhost:4000/graphql ./src"
```

9. In a separate terminal, run your backend with **MongoDB**, plus the Docker container from *Chapter 8* (it should be running on port 4000). You can also run the backend from the previous chapter from the book's repository.

10. In another terminal, run the npm zeus command:

```
npm run zeus
```

This command will connect to our GraphQL Server and generate TypeScript type definitions out of it.

11. Install **React Router**, which will handle all the navigation in our project:

```
npm install react-router-dom
```

12. Install jotai, which will be used to manage the global state of the application:

```
npm i jotai
```

13. Install prettier and the eslint plugin, which we will use to format our code:

```
npm i -D prettier eslint-config-prettier eslint-plugin-prettier
```

14. Create a .prettierrc file in the root of the project and add the following content:

```
{
    "trailingComma": "all",
    "tabWidth": 2,
    "semi": true,
    "singleQuote": true,
    "printWidth": 120,
    "bracketSpacing": true,
    "endOfLine": "lf",
    "plugins": [
```

```
                    "prettier-plugin-tailwindcss"
        ]
    }
```

15. Add a `prettier` plugin to your `.eslintrc.cjs` file. This will allow the linter to communicate with the formatter. The file should look like this:

```
module.exports = {
  root: true,
  env: { browser: true, es2020: true },
  extends: [
    'eslint:recommended',
    'plugin:@typescript-eslint/recommended',
    'plugin:react-hooks/recommended',
    "plugin:prettier/recommended"
  ],
  ignorePatterns: ['dist', '.eslintrc.cjs'],
  parser: '@typescript-eslint/parser',
  plugins: ['react-refresh'],
  rules: {
    'react-refresh/only-export-components': [
      'warn',
      { allowConstantExport: true },
    ],
  },
}
```

Although the setup process was quite lengthy, it will save us time when working on the frontend of our solution. In the next section, we will focus on properly creating the GraphQL client.

Preparing our GraphQL Layer

Before we start creating components and styling them, we need to create a good communication layer between our backend and frontend. We will write selectors and prepare the client to communicate with the backend in two states – authorized and not authorized.

Writing selectors

We will begin by creating a `graphql` folder inside our project's `/src` folder and adding a `selectors.ts` file. Inside this file, we will write **selectors** for our GraphQL queries, and the types that will be used to type the application state.

First, what is a selector? A selector is Zeus' version of a GraphQL fragment. We discussed fragments in *Chapter 1*, so go back for a recap.

Second, why do we need selectors? They can be used to create TypeScript types needed by React state and help persist type-safe data there.

The primary selector utilized for the queries on our system's **Home** page is based on the QuestionsResponse type derived from the GraphQL schema discussed in the previous chapter. This selector plays a crucial role in fetching and filtering the necessary data from the QuestionsResponse type, allowing us to efficiently retrieve and display the relevant information on the home page. We can see the selector here:

```
import { FromSelector, Selector } from "@/zeus";

const QuestionBaseSelector = Selector("Question")({
  _id: true,
  score: true,
  title: true,
  createdAt: true,
  user: {
    username: true,
  },
});

const AnswerBaseSelector = Selector("Answer")({
  _id: true,
  content: true,
  score: true,
  createdAt: true,
  user: {
    username: true,
  },
});

export const AnswerDetailSelector = Selector("Answer")({
  ...AnswerBaseSelector,
  answers: AnswerBaseSelector,
});

export const QuestionDetailSelector =
  Selector("Question")
({
  ...QuestionBaseSelector,
  content: true,
  answers: AnswerBaseSelector,
});
export type QuestionDetailType = FromSelector<
```

```
      typeof QuestionDetailSelector,
      "Question"
  >;

  export const QuestionResponseListSelector =
    Selector("QuestionsResponse")
  ({
    question: QuestionBaseSelector,
    bestAnswer: AnswerBaseSelector,
  });

  export type QuestionResponseListType = FromSelector<
      typeof QuestionResponseListSelector,
      "QuestionsResponse"
  >;
```

`QuestionResponseListSelector` is written using Zeus-generated code to achieve TypeScript compatibility for type safety on the frontend. Every field inside the selector that has the `true` value is the equivalent of that field in the GraphQL Query Language.

Then `QuestionResponseListType` is created out of the selector to be used as a type for variables created by the `useState` hook.

Both the selector and type will be used for performing the GraphQL two root queries: `search` and `topQuestions`.

Now, to better understand how selectors are similar to fragments, let's see how the query generated from the selector looks in GraphQL Query Language:

```
query top {
  top {
    question {
      score
      title
      _id
      createdAt
      user {
        username
      }
    }
    bestAnswer {
      content
      score
      user {
        username
```

```
          }
        }
      }
    }
```

This query was generated from our selector. With that, you can see how the GraphQL Zeus library generates GraphQL Query Language code. Later in this chapter, you will learn more about this library, and use the selectors when we query the backend.

After preparing the selectors, we need to prepare the client library supporting authorization and authentication.

Implementing a client hook

To execute all the queries with the selector, we need a simple client hook to hold our authorization data. **React hooks** are a powerful feature introduced in React 16.8 that allows developers to use state and other React features in functional components. With hooks, developers can easily manage state, handle side effects, and reuse logic across components

We will put the client hook inside the `src/graphql/client.ts` file, side by side with our selector's code, like so:

```
import { jwtToken } from "@/atoms";
import { Chain, HOST } from "@/zeus";
import { useAtom } from "jotai";
import { useMemo } from "react";

export const useClient = () => {
  const [token, setToken] = useAtom(jwtToken);
  const client = useMemo(() => {
    return Chain(HOST, {
      headers: {
        "Content-Type": "application/json",
        ...(token
          ? {
              Authorization: token,
            }
          : {}),
      },
    });
  }, [token]);
  const login = useCallback(
    (username: string, password: string) => {
      client('mutation')({
```

```
      public: {
        login: [
          { username, password },
          {
            token: true,
          },
        ],
      },
    }).then((response) => {
      const token = response.public?.login.token;
      if (token) {
        setToken(token);
      }
    });
  },
  [client, setToken],
);

const register = useCallback(
  (username: string, password: string) => {
    client('mutation')({
      public: {
        register: [
          { username, password },
          {
            token: true,
          },
        ],
      },
    }).then((response) => {
      const token = response.public?.register.token;
      if (token) {
        setToken(token);
      }
    });
  },
  [client, setToken],
);

return {
  client,
  setToken,
  login,
```

```
    register,
    isLoggedIn: !!token,
  };
};
```

Breaking down all of that code, first of all, we import jwtToken from an **atom**. Here, we are referring to a jotai atom from the src/atoms.ts file, which looks like this:

```
import { atom } from "jotai";

export const jwtToken = atom("token", "");
```

An atom is responsible for holding a variable's value and re-rendering the component using the atom every time the value of the atom changes.

Next, we import the Chain function, which is a type-safe fetch wrapper for GraphQL backends, and a HOST constant, which is the same host we provided in the Zeus script inside package.json.

Inside the useClient hook, we consume the jwtToken atom from jotai to allow the ability to read and set a value on the token. Then we memoize the client based on that token to avoid unnecessary re-initialization. Before the token is set – meaning before the user is logged in – we return the unauthorized client, which transforms into the authorized client after the user logs in.

We then implemented the login and register functions to call our backend mutations and to store the token after successfully logging in or registering.

In the end, we return the client, the setToken function, and the isLoggedIn helper variable from our React hook.

Believe it or not, we are now prepared to implement our frontend!

Creating the system's frontend

As we have prepared the client and the GraphQL layer together with selectors and TypeScript types, we can now code the components, routing, and application logic. In this section, we will set up the frontend and make a publicly visible **Home** page.

Setting up the frontend

In this project, we will mostly use the **Shadcn UI** for the visual part of our frontend. It looks nice, doesn't need any additional styling work for its components, and offers full customizability.

First of all, we can generate components for our UI (later on, we can use those components to compose full layouts out of them):

```
npx shadcn-ui@latest add input button label card textarea sonner
```

Next, we need to set up the main app file and the routing structure. So, inside the `src/main.tsx` file, insert the following content:

```
import React from "react";
import ReactDOM from "react-dom/client";
import App from "./App.tsx";
import "./index.css";

ReactDOM.createRoot(
  document.getElementById("root")!
).render(
  <React.StrictMode>
    <App />
  </React.StrictMode>,
);
```

Inside the main app file, we render the React app into the HTML div with the `root` ID.

We also import `index.css` generated by Tailwind at the top of the file, which we need for the `tailwind` library to work with our project.

Tailwind needs the directives to be present inside the CSS file. Here is the content of our `index.css` file:

```
@tailwind base;
@tailwind components;
@tailwind utilities;

#root,body,html{
    width: 100%;
    height: 100%;
}
```

Here, we can observe the Tailwind directives that import all the Tailwind features into our project. We also set the full width and height of our main components.

Next, inside `src/App.tsx` (which we imported inside `src/main.tsx`), we will create a routing tree for our `react-router` package and return the `react-router` app:

```
import {
  createBrowserRouter,
  RouterProvider
} from "react-router-dom";
import Home from "@/Home";
import Root from "@/Root";
import AuthRoot from "@/auth/AuthRoot";
```

```
import Login from "@/auth/Login";
import Register from "@/auth/Register";
import AuthGuard from "@/authorized/AuthGuard";
import PostQuestion from "@/authorized/PostQuestion";
import Question from "@/pages/Question";

const router = createBrowserRouter([
  {
    path: "",
    element: <Root />,
    children: [
      {
        path: "",
        element: <Home />,
      },
      {
        path: "question/:questionId",
        element: <Question />,
      },
      {
        path: "auth",
        element: <AuthRoot />,
        children: [
          {
            path: "login",
            element: <Login />,
          },
          {
            path: "sign-up",
            element: <Register />,
          },
        ],
      },
      {
        path: "me",
        element: <AuthGuard />,
        children: [
          {
            path: "post",
            element: <PostQuestion />,
          },
        ],
      },
```

```
    ],
  },
]);

function App() {
  return <RouterProvider router={router} />;
}


export default App;
```

Inside the tree, we specified the following:

- Every page in the tree should use the <Root /> component as the layout. This means that it will render the <Root /> component and render the children inside the <Outlet /> component that is included in the root layout.

- Every page where the URL begins with auth should also use the <AuthRoot /> layout.

- Every page where the URL begins with me should also use the <AuthGuard /> layout.

- Questions should pass parameters to the browser when requested by its ID; for example, /question/2jn139i8h298713h should pass the questionId parameter to the rendered page with the 2jn139i8h298713h value.

Okay, we have the routing tree. Now we can take a look at our <Root /> layout that will be used all across the app. Its purpose is to hold the components that will be shared across pages that are children of it. Here is the code of the Root layout:

```
import { jwtToken } from "@/atoms";
import { Button } from "@/components/ui/button";
import { useAtom } from "jotai";
import { Link, Outlet } from "react-router-dom";
import { Toaster } from '@/components/ui/sonner';

const Root = () => {
  const [token, setToken] = useAtom(jwtToken);
  return (
    <>
      <nav
        className=
          "w-full absolute hidden flex-col gap-6 text-lg
          font-medium md:flex md:flex-row md:items-center
          md:gap-5 md:text-sm lg:gap-6"
      >
        <div
          className=
```

```
          "container flex items-center w-full"
      >
        <Link
          to="/"
          className="text-muted-foreground
          transition-colors hover:text-foreground px-8
          py-4 block"
        >
          Home
        </Link>
        <div
          className=
            "ml-auto space-x-2 flex items-center"
        >
          {!!token && (
            <>
              <p>User logged in</p>
              <Link to="/me/post">
                <Button>Post question</Button>
              </Link>
              <Button
                variant={"outline"}
                onClick={() => {
                  setToken(null);
                }}
              >
                Logout
              </Button>
            </>
          )}
          {!token && (
            <>
              <Link to="/auth/login">
                <Button>Login</Button>
              </Link>
              <Link to="/auth/register">
                <Button
                  variant={"outline"}
                >
                  Register
                </Button>
              </Link>
            </>
```

```
            )}
          </div>
        </div>
      </nav>
      <div
        className=
          "flex w-full gap-4 md:ml-auto md:gap-2 lg:gap-4
          min-h-full pt-8 container"
      >
        <Outlet />
      </div>
      <Toaster />
    </>
  );
};
export default Root;
```

Inside our `Root` layout, we consume the `jwtToken` value – this will be used to determine whether a user is logged in or not. Then, we have the component that points to the **Home** page. Next, depending on whether the token is set, we return the **Login** and **Sign-up** buttons or the **Logout** and **Post Question** buttons. In the end, we have the `Outlet` component, which is used to render the Root layout's current active children within the layout.

At this point, we have the `Root` layout ready, and `useClient` is done. Now we can code the **Home** page.

Making the Home Page

Besides the navigation bar, our **Home** page will contain a search button, along with the top ten questions with the highest scores out of all questions displayed together, and a shortened best answer to attract the user.

Before we start composing the page, we need some components. The first is the `QuestionList` component that will be used on the **Home** page inside `src/pages/Home/QuestionList.tsx`:

```
import {
  QuestionTile
} from '@/components/ui/molecules/question';
import {
  QuestionResponseListType
} from '@/graphql/selectors';

export const QuestionList = ({
  questions,
  title,
```

```
    }: {
      questions: QuestionResponseListType[];
      title: string;
    }) => {
      return (
        <div className="space-y-4 flex flex-col">
          <h1
            className=
              "scroll-m-20 text-2xl font-bold text-gray-600
              tracking-tight lg:text-3xl"
          >
            {title}
          </h1>
          {questions.map((questionResponse) => {
            return (
              <QuestionTile
                questionResponse={questionResponse}
                key={questionResponse.question._id}
              />
            );
          })}
        </div>
      );
    };
```

Inside this `QuestionList` component, we take `questions` and `title` as parameters of the function and display them in a list of individual `QuestionTile` components.

Inside the `QuestionTile` component, we need to include the following information returned from the backend:

- Who posted the question and when.

- The title as a short form of the question.

- A shortened best answer content to attract users.

- The `RocketButton` component for voting on the question. We will have one voting button, which we will use in places where we display the question or answer score. Clicking this button will trigger the vote mutation.

We can include all those points inside the `QuestionTile` component by entering the following code inside `src/components/molecules/question.tsx`:

```
import {
  RocketButton
```

```
} from '@/components/ui/atoms/rocketbutton';
import {
  QuestionResponseListType
} from '@/graphql/selectors';
import { Link } from 'react-router-dom';

export const QuestionTile = ({
  questionResponse: { question, bestAnswer },
}: {
  questionResponse: QuestionResponseListType;
}) => {
  return (
    <div className="border rounded p-4 max-w-md">
      <Link to={`/question/${question._id}`}>
        <div className="space-x-4 ">
          <div>
            <i className="text-gray-400">
              {`${new Date(question.createdAt)
              .toLocaleString()} ${question.user.username}
              asked`}
            </i>
            <div className="text-xl">{question.title}</div>
            {bestAnswer && (
              <>
                <b>Best answer</b>
                <p>
                  {bestAnswer.content
                    .split('s')
                    .slice(0, 10)
                    .join('s')}
                  ...
                </p>
              </>
            )}
          </div>
        </div>
      </Link>
      <div className="justify-end flex">
        <RocketButton {...question} />
      </div>
    </div>
  );
};
```

Here, we made good use of `QuestionResponseListType`, providing a type for the properties of this `QuestionTile` component. This way, we can just pass the object returned from the backend to the component.

> **Note**
>
> I am not a big fan of the *Don't Repeat Yourself* rule. However, in this case, with the selectors and the types inferred from the generated code, we avoided writing separate types for the frontend and inherited them from the backend system instead.

Our `QuestionTile` component uses `QuestionResponseListType`, which is returned from the backend and displays the question together with the most-voted answer based on that data. It also includes the `RocketButton` component that, as mentioned, will be utilized throughout our application to allow users to vote for both questions and answers. Additionally, the list tile links to a detailed view of the question, displaying a list of answers. Here is our main `RocketButton` implementation:

```
import { Button } from '@/components/ui/button';
import { useClient } from '@/graphql/client';
import { Rocket } from 'lucide-react';
import { useState } from 'react';
import {
  useLocation,
  useNavigate
} from 'react-router-dom';

export const RocketButton = ({
  score,
  _id,
}: {
  score: number;
  _id: string;
}) => {
  const [currentScore, setCurrentScore] = useState(score);
  const { client, isLoggedIn } = useClient();
  const nav = useNavigate();
  const location = useLocation();
  return (
    <Button
      variant={'outline'}
      title={`${currentScore} rockets received. Click to
        give some rockets`}
      onClick={(e) => {
        e.stopPropagation();
```

```
      if (!isLoggedIn) {
        nav(`/auth/login?next=${location.pathname}`);
        return;
      }
      client('mutation')({
        user: {
          vote: [{ _id: _id }, true],
        },
      }).then((r) => {
        if (r.user?.vote) {
          setCurrentScore((cs) => cs + 1);
          return;
        }
      });
    }}
  >
    <Rocket className="mr-2" />
    <p>{currentScore}</p>
  </Button>
  );
};
```

We added a rocket icon from the `lucide-react` icons and included information about the question or answer's score. After clicking the button, it will update the state of the score locally and send a mutation to our API to vote for a question to increase its score to the backend. If the user is not logged in, they will be automatically redirected to the login page. This redirection will include the current location and pass the current URL. This allows the user to be redirected back to the specific page they intended to access after a successful login.

Now, our task is to create a hook that will handle the queries for the **Home** page. This hook should fetch the top questions and also incorporate the search functionality that allows users to search for specific content directly on the home page. Here is the code that uses our GraphQL client inside `src/Home/useHomeQueries.ts`:

```
import { useClient } from '@/graphql/client';
import {
  AnswerDetailSelector,
  QuestionDetailSelector,
  QuestionResponseListSelector,
  QuestionResponseListType,
} from '@/graphql/selectors';
import { useCallback, useEffect, useState } from 'react';

export const useHomeQueries = () => {
```

```
const [topQuestions, setTopQuestions] =
  useState<QuestionResponseListType[]>();
const [foundQuestions, setFoundQuestions] =
  useState<QuestionResponseListType[]>();
const [submitValue, setSubmitValue] = useState('');
const { client } = useClient();
useEffect(() => {
  client('query')({
    top: QuestionResponseListSelector,
  }).then((r) => {
    setTopQuestions(r.top);
  });
  // eslint-disable-next-line react-hooks/exhaustive-deps
}, []);
const search = useCallback(
  (q: string) => {
    setSubmitValue(q);
    client('query')({
      search: [
        { query: q },
        {
          question: QuestionDetailSelector,
          bestAnswer: AnswerDetailSelector,
        },
      ],
    }).then((r) => {
      setFoundQuestions(r.search);
    });
  },
  [client],
);
return {
  search,
  questions: foundQuestions
    ? {
        questions: foundQuestions,
        title: `Search results for "${submitValue}"`,
      }
    : {
        questions: topQuestions,
        title: 'Top questions from the community',
      },
```

```
      submitValue,
  };
};
```

First, we used the `useState` hook to hold values of the `QuestionResponseListType[]` type returned from our GraphQL backend.

Then, we wrote the `useEffect` hook to fetch top questions when the hook is loaded. We didn't include `client` as a hook dependency as it doesn't matter whether the user is authenticated. Questions returned from this `topQuestions` query are inserted into the `topQuestions` state.

We also coded the `search` function and wrapped it into `useCallback` – this will search for questions and insert the returned questions into the `foundQuestions` state.

Finally, we returned the `search` function and the state values from our hook.

Now our hook is ready to be used in the **Home** page component. Here is the awaited **Home** page component utilizing all the prepared components and queries:

```
import { Button } from '@/components/ui/button';
import { Input } from '@/components/ui/input';
import { useState } from 'react';
import { Link } from 'react-router-dom';
import { QuestionList } from '@/pages/Home/QuestionList';
import {
  useHomeQueries
} from '@/pages/Home/useHomeQueries';

const Home = () => {
  const { questions, search, submitValue } =
    useHomeQueries();
  const [searchValue, setSearchValue] = useState('');
  return (
    <div
      className=
        "flex flex-col space-y-16 container items-center
        pb-24"
    >
      <div
        className=
          "space-y-4 flex flex-col w-full items-center
          pt-[20vh]"
      >
        <h1
          className=
```

```
          "scroll-m-20 text-4xl font-extrabold
          tracking-tight lg:text-5xl"
      >
        Find an answer
      </h1>
      <div
        className=
          "flex w-full max-w-md items-center space-x-2"
      >
        <Input
          value={searchValue}
          onChange={(e) =>
            setSearchValue(e.target.value)}
          type="text"
          placeholder="Search for question"
        />
        <Button
          onClick={() => {
            search(searchValue);
          }}
          type="submit"
        >
          Search
        </Button>
      </div>
    </div>
    {questions.questions?.length && (
      <QuestionList
        questions={questions.questions}
        title={questions.title}
      />
    )}
    {questions.questions?.length === 0 && (
      <div
        className=
          "flex flex-col space-y-2 max-w-md items-end"
      >
        <p className="text-gray-500">
          {`No results for "${submitValue}". You can
          create your question and post it here! `}
        </p>
        <Link to="/me/post">
          <Button variant={'secondary'}>
```

```
            Post question
          </Button>
        </Link>
      </div>
    )}
  </div>
  );
};
export default Home;
```

We started by using our `useHomeQueries` hook to get the data. Then we implemented `searchValue` to hold the value of the search input. Next, we created the container layout for the **Home** page. After that, inside the container, we displayed the **Find an answer** heading, text input with a search button to be used for search functionality, and a list of top-rated questions.

Moving on, we implemented an empty state when the search results are `empty`. Inside this empty state, we added a button that takes the user to the **Post question** page (but before we actually allow that, we need to implement the **Login and Registration** frontend).

This design pattern is called **First-Time User Experience** (**FTUX**), which refers to the initial interaction a user has with a product or application when they use it for the first time. It focuses on creating a positive and intuitive experience, guiding users through the key features and functionalities. A well-designed FTUX ensures that users can quickly understand and navigate the product, leading to higher user satisfaction and increased engagement. In our app, we will use this pattern in the empty component with the **Post question** button that triggers user action.

At this point, our **Home** page should look like this:

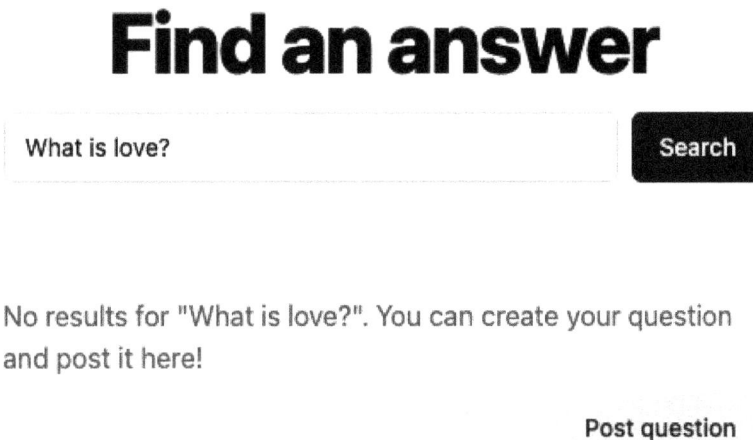

Find an answer

| What is love? | **Search** |

No results for "What is love?". You can create your question and post it here!

Post question

Figure 9.1: Home page of our service

We have successfully implemented a clean-looking layout for the **Home** page of our Questions and Answers service. Not only did we use React components but we also played with hooks and generated client functions. Now we need to allow users to create accounts.

Creating authentication pages

To allow users to post and reply to questions, we need to provide a way for them to register and log in. As stated in the `Root` layout, those pages will have the `src/pages/auth/AuthRoot.tsx` layout as their layout component.

Inside the layout for authentication pages, we need to redirect users back to the **Home** page if they are already logged in. We don't want to display the register and login forms for already logged-in users. This feature becomes particularly useful when the browser remembers the login link with the next parameter. In such cases, instead of encountering an error, users can be seamlessly redirected to the appropriate page.

Here is the `AuthRoot` component code that is the base for the login and register forms:

```
import { jwtToken } from '@/atoms';
import { useAtom } from 'jotai';
import { useEffect } from 'react';
import {
  Outlet,
  useNavigate,
  useSearchParams
} from 'react-router-dom';
const AuthRoot = () => {
  const [token] = useAtom(jwtToken);
  const [searchParams] = useSearchParams();
  const n = useNavigate();
  useEffect(() => {
    if (token) {
      const hasNext = searchParams.get('next');
      if (hasNext) {
        n(hasNext);
        return;
      }
      n('/');
      return;
    }
  }, [token, searchParams, n]);

  return (
    <div
```

```
      className=
        "flex items-center justify-center w-full"
    >
      <Outlet />
    </div>
  );
};
export default AuthRoot;
```

Our `AuthRoot` layout contains one `useEffect` hook, which will be triggered only when the token is set. The navigation function, `n`, will then redirect the user to the page specified in the `next` query parameter, or to the **Home** page when the logged-in user enters the `auth/login` URL path directly. The `<Outlet />` component then renders the children of this `AuthRoot` component defined in the `react router` setup.

Now we will take a look at the actual **Sign Up** page:

```
import { Button } from '@/components/ui/button';
import {
  Card,
  CardContent,
  CardDescription,
  CardHeader,
  CardTitle,
} from '@/components/ui/card';
import { Input } from '@/components/ui/input';
import { Label } from '@/components/ui/label';
import { useClient } from '@/graphql/client';
import { useState } from 'react';
import { Link } from 'react-router-dom';

const Register = () => {
  const [username, setUsername] = useState('');
  const [password, setPassword] = useState('');
  const { register } = useClient();
  return (
    <div>
      <Card className="mx-auto max-w-sm">
        <CardHeader>
          <CardTitle className="text-xl">
            Sign Up
          </CardTitle>
          <CardDescription>
            Enter your information to create an account
```

```
        </CardDescription>
      </CardHeader>
    <CardContent>
      <div className="grid gap-4">
        <div className="grid gap-2">
          <Label htmlFor="email">Email</Label>
          <Input
            id="email"
            type="email"
            placeholder="m@example.com"
            required
            value={username}
            onChange={(e) =>
              setUsername(e.target.value)}
          />
        </div>
        <div className="grid gap-2">
          <Label htmlFor="password">Password</Label>
          <Input
            value={password}
            onChange={(e) =>
              setPassword(e.target.value)}
            id="password"
            type="password"
          />
        </div>
        <Button
          type="submit"
          className="w-full"
          onClick={() => {
            if (username && password)
              register(username, password);
          }}
        >
          Create an account
        </Button>
      </div>
      <div className="mt-4 text-center text-sm">
        Already have an account?{' '}
        <Link to="/auth/login" className="underline">
          Sign in
        </Link>
      </div>
```

```
        </CardContent>
      </Card>
    </div>
  );
};

export default Register;
```

To make this page look nice, we used the `Card` component. For the upper part of the card, we used `CardHeader` with `CardTitle` and `CardDescription`, and for the bottom part, we used `CardContent`. We also used the `useState` function to hold the `username` and `password` values and the `register` function from our `useClient` hook.

Here is the result of the **Sign Up** page:

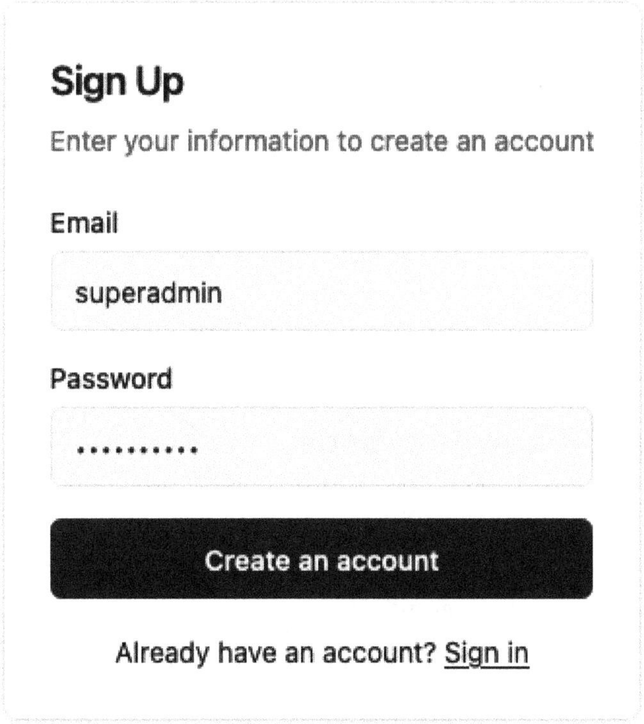

Figure 9.2: Sign Up form

The **Login** page looks almost the same, and to save us from repeating ourselves, the code is available in the chapter's GitHub folder. The only difference is the text that the page contains and that we used the `login` function from `useClient` instead of `register`. Here is what the **Login** page looks like:

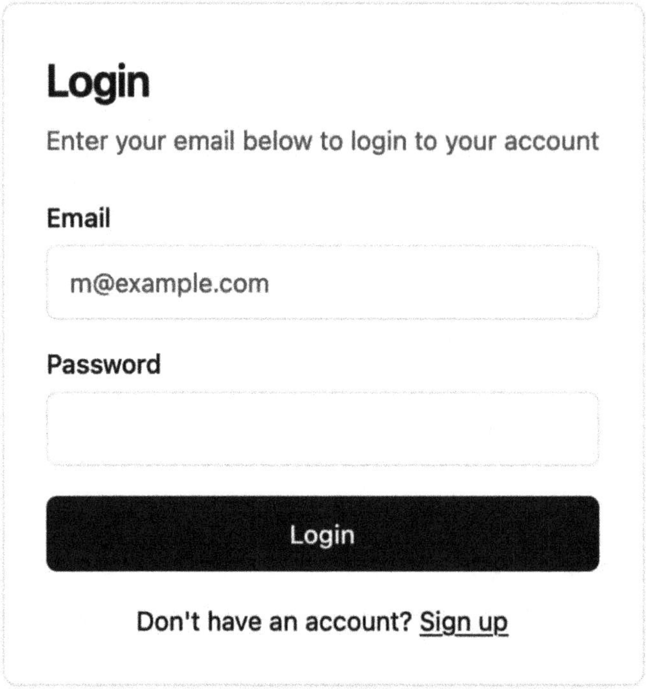

Figure 9.3: Login form

If everything works properly, we can log in and create a **Post your question** page. Let's start creating that page next.

Crafting the post question page

To create pages that need users to be logged in, we need to create a separate folder with a dedicated guard layout component. A **guard layout component** is a piece of code that prevents unauthorized users from entering certain parts of the frontend. Without proper authorization and guard layouts, the backend functionality on certain pages will not be accessible, resulting in a poor user experience. So, inside the guard layout component, we check whether the user is logged in, and if they are not, we redirect them to the login form.

In this layout, we will try to redirect users to the exact page they want after the authentication process. We will put this layout inside src/pages/me/AuthGuard.tsx:

```
import { jwtToken } from '@/atoms';
import { useAtom } from 'jotai';
import { useEffect } from 'react';
import {
  Outlet,
```

```
    useLocation,
    useNavigate
} from 'react-router-dom';
const AuthGuard = () => {
  const [token] = useAtom(jwtToken);
  const nav = useNavigate();
  const location = useLocation();
  useEffect(() => {
    if (!token) {
      nav('/auth/login?' + `next=${location.pathname}`);
      return;
    }
  }, [token, nav, location.pathname]);
  return (
    <div
      className=
        "w-full justify-center items-center flex
        flex-col container space-y-8"
    >
      <Outlet />
    </div>
  );
};
export default AuthGuard;
```

In this `AuthGuard` guard layout component, we will also use the `useEffect` function that will automatically redirect the user to the authentication **Login** page if they try to enter any layout children pages without being logged in. Then, if the user is logged in, we do nothing and display the children component.

Next, we need to prepare the hook with queries for the authorized user. We will put this file inside `src/pages/me/useMeQueries.ts`, along with the following code:

```
import { useClient } from '@/graphql/client';
import { useCallback } from 'react';

export const useMeQueries = () => {
  const { client } = useClient();

  const postQuestion = useCallback(
    (title: string, description: string) => {
      return client('mutation')({
        user: {
          postQuestion: [
```

```
              {
                createQuestion: {
                  content: description,
                  title,
                },
              },
              true,
            ],
          },
        });
      },
    [client],
  );


  return { postQuestion };
};
```

In this file, we have only one `postQuestion` function, which is a mutation responsible for posting questions by the user.

We are now prepared to create the **Post your question** page. We will do this inside `src/pages/me/PostQuestion.tsx`, and it will contain a form to post a question to our service. Here is the content of the file:

```
import { Button } from '@/components/ui/button';
import {
  Card,
  CardContent,
  CardDescription,
  CardHeader,
  CardTitle,
} from '@/components/ui/card';
import { Input } from '@/components/ui/input';
import { Textarea } from '@/components/ui/textarea';
import { useMeQueries } from '@/pages/me/useMeQueries';
import { useState } from 'react';
import { useNavigate } from 'react-router-dom';
import { toast } from 'sonner';

const PostQuestion = () => {
  const [title, setTitle] = useState('');
  const [description, setDescription] = useState('');
  const { postQuestion } = useMeQueries();
  const navigate = useNavigate();
```

```
return (
  <>
    <Card className="mx-auto max-w-md">
      <CardHeader>
        <CardTitle className="text-2xl">
          Post your question!
        </CardTitle>
        <CardDescription>
          Post your question and wait for answers
        </CardDescription>
      </CardHeader>
      <CardContent>
        <div className="space-y-4">
          <Input
            required
            placeholder="Question title"
            value={title}
            onChange={(e) => setTitle(e.target.value)}
          />
          <Textarea
            placeholder="Question longer content"
            required
            value={description}
            onChange={(e) =>
              setDescription(e.target.value)}
          />
          <div>

            <Button
              onClick={() =>
                postQuestion(title, description)
                  .then(() => {
                    toast('Successfully posted
                      question');
                    navigate('/');
                  })
              }
            >
              Post question
            </Button>
          </div>
        </div>
```

```
      </CardContent>
    </Card>
  </>
 );
};
export default PostQuestion;
```

Here, we have a form using the `Input` and `TextArea` components. We hold that form inside the `Card` component with the title and description. Then, after clicking the **Post question** button, the `postQuestion` function is fired using the data from our form.

Here is the result of the **Post your question** page:

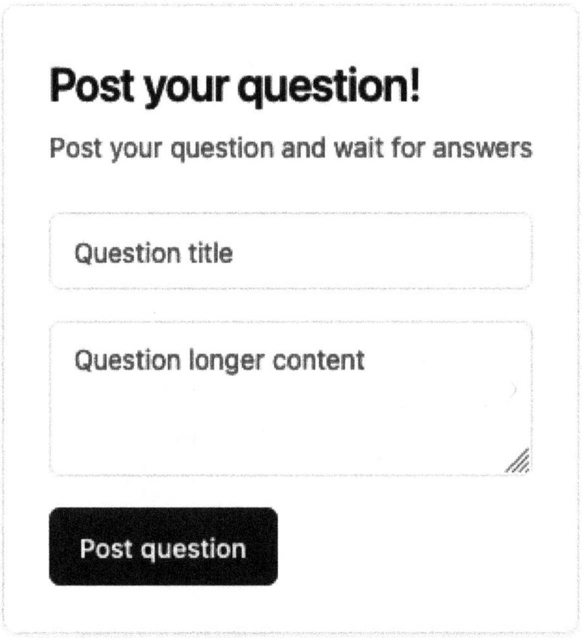

Figure 9.4: Post your question form

Now we have created the possibility for users to post questions, as well as a guard layout to redirect users who are not logged in to authentication pages. Next, we need to implement a page where users can view more details about the questions and answer them.

Implementing the questions page

The last page we will create is the **Question** page – this page will display the question along with the answers.

Again, before we start making components, we have to prepare the functions that call the GraphQL server inside our custom hook – just like we did for the **Home** page. To do that, let's create a `src/pages/Question/useQuestionQuery.ts` file with the following contents:

```
import { useClient } from '@/graphql/client';
import {
  QuestionDetailSelector,
  QuestionDetailType,
} from '@/graphql/selectors';
import { useCallback, useEffect, useState } from 'react';

export const useQuestion = (questionId:
  string | undefined
) => {
  const [loadingState, setLoadingState] = useState<
    'loading' | '404' | 'loaded'
  >('loading');
  const [currentQuestion, setCurrentQuestion] =
    useState<QuestionDetailType>();
  const { client } = useClient();

  const fetchQuestion = useCallback(
    (questionId: string) => {
      if (!questionId) {
        setLoadingState('404');
        return;
      }
      client('query')({
        question: [
          { _id: questionId },
          QuestionDetailSelector
        ],
      })
        .then((q) => {
          setCurrentQuestion(q.question);
          if (!q.question) setLoadingState('404');
          else setLoadingState('loaded');
        })
        .catch(() => {
          setLoadingState('404');
        });
    },
    [client],
  );
```

```
const answer = useCallback(
  (
    questionId: string,
    answeringToId: string,
    text: string
  ) => {
    client('mutation')({
      user: {
        postAnswer: [
          {
            createAnswer: {
              content: text,
              to: answeringToId,
            },
          },
          true,
        ],
      },
    }).then((r) => {
      if (r.user?.postAnswer) {
        fetchQuestion(questionId);
      }
    });
  },
  [client, fetchQuestion],
);

useEffect(() => {
  if (questionId) {
    fetchQuestion(questionId);
  } else {
    setLoadingState('404');
  }
}, [questionId, fetchQuestion]);

return { answer, currentQuestion, loadingState };
};
```

At the start of the useQuestion hook function, we have loadingState, which holds the information about the current state of query execution.

Next, we have currentQuestion with QuestionDetailType from our selectors, which will hold the current question.

After that, the `fetchQuestion` function gets the question from the backend with the answers to this question. If the question with the `questionId` exists, we map the returned value to the `currentQuestion` state variable. If the question does not exist, we change the state to `'404'`.

Then, in the answer function, we call the `postAnswer` mutation with the desired data. If the answer action is successful, we refresh the current question to display our answer inside it.

At the end of the `useQuestion` hook function, we have the `useEffect` hook that fetches the question based on the `questionId` parameter. Every time `questionId` changes, it will re-fetch the current question details.

With our hook function ready, we can move on and implement the React component. On the **Question** page, we will have the question, the answers, and the possibility to answer the question. Here is the code for that page:

```
import {
  RocketButton
} from '@/components/ui/atoms/rocketbutton';
import { Button } from '@/components/ui/button';
import { AnswerForm } from '@/pages/Question/AnswerForm';
import { AnswerTile } from '@/pages/Question/AnswerTile';
import { Question404 } from '@/pages/Question/Question404';
import {
  useQuestion
} from '@/pages/Question/useQuestionQuery';
import { useState } from 'react';
import { useParams } from 'react-router-dom';

const Question = () => {
  const params = useParams();
  const [answeringToId, setAnsweringToId] =
    useState<string>();
  const questionId =
    params.questionId as string | undefined;
  const { currentQuestion, loadingState } =
    useQuestion(questionId);
  if (loadingState === 'loading') {
    return (
      <div
        className=
          "w-full flex flex-col max-w-md m-auto space-y-4"
      >
        <p className="text-center">
          Loading... please wait
        </p>
```

```
      </div>
    );
  }
  if (loadingState === '404' || !currentQuestion) {
    return <Question404 questionId=
      {questionId || 'no question id'} />;
  }
  return (
    <div
      className=
        "w-full flex flex-col max-w-md m-auto space-y-4
        pt-[20vh] pb-24"
    >
      <div className="flex items-center space-x-2">
        <RocketButton {...currentQuestion} />
        <i>
          On {new Date(currentQuestion.createdAt)
            .toLocaleString()}{' '}
          {currentQuestion.user.username} asked
        </i>
      </div>
      <h2 className="font-extrabold text-3xl">
        {currentQuestion?.title}
      </h2>
      <p>{currentQuestion?.content}</p>
      {answeringToId !== currentQuestion._id && (
        <div className="flex justify-end">
          <Button
            onClick={() =>
              setAnsweringToId(currentQuestion._id)
            }
          >
            Answer
          </Button>
        </div>
      )}
      {answeringToId === currentQuestion._id && (
        <AnswerForm
          questionId={currentQuestion._id}
          answerToId={answeringToId}
          cancel={() => setAnsweringToId(undefined)}
        />
      )}
```

```
        <h3 className="text-lg border-b">Top Answers</h3>
        <div className="flex flex-col space-y-4">
          {currentQuestion.answers.map((a) => (
            <AnswerTile key={a._id} answer={a} />
          ))}
        </div>
      </div>
    );
  };


  export default Question;
```

Using the useParams hook from react-router, we started by gathering URL parameters to check what the actual questionId query parameter is inside our URL path. Then we extracted the questionId parameter from the URL – we need to remember that the questionId parameter inside the URL can point to a question that doesn't exist anymore. This parameter comes from the URL bar so it can be stored by the browser or entered manually by users.

Next, we used our useQuestion hook to get the current question and its loading state. If the question is still loading, we display the loading text – **Loading… please wait** – and if the ID is wrong, we display the Question404 component.

Then, inside the returned **JavaScript XML (JSX)** TypeScript code, we display the question's details, including the date of creation and the username of the posting user.

> **Note**
>
> JSX is a syntax extension used in React and other JavaScript frameworks to define and structure UI components in a declarative manner, combining HTML-like syntax with JavaScript logic.

Finally, under the displayed question, we display the **Answer** button. If we click the button, the AnswerForm component will appear allowing us to answer the question.

Let's code this answer form as a separate component in ./src/pages/Question/AnswerForm. tsx:

```
import { Button } from '@/components/ui/button';
import { Textarea } from '@/components/ui/textarea';
import {
  useQuestion
} from '@/pages/Question/useQuestionQuery';
import {
  useEffect,
  useState
```

```
} from 'react';

export const AnswerForm = ({
  questionId,
  answerToId,
  cancel,
}: {
  questionId: string;
  answerToId: string;
  cancel: () => void;
}) => {
  const [answerContent, setAnswerContent] = useState('');
  const { answer } = useQuestion(questionId);
  useEffect(() => {
    setAnswerContent('');
  }, [answerToId]);
  return (
    <div
      className=
        "w-full flex flex-col max-w-md m-auto space-y-4
        pt-[20vh] pb-24"
    >
      <div className="flex flex-col space-y-4">
        <Textarea
          placeholder="write"
          value={answerContent}
          onChange={(e) =>
            setAnswerContent(e.target.value)}
        />
        <div className="flex space-x-4 justify-end">
          <Button
            onClick={() => cancel()}
            variant={'ghost'}
          >
            Cancel
          </Button>
          <Button
            onClick={() => {
              answer(questionId, answerToId,
              answerContent);
            }}
          >
            Answer
```

```
          </Button>
        </div>
      </div>
    </div>
  );
};
```

Inside the `AnswerForm` component, we have the form that allows the user to answer the question. In the beginning, we have the `answerContent` state variable. We then clear this variable value every time the question or answer ID changes inside the `useEffect` hook. Then, inside the JSX `return` statement, we have the `Textarea` component and a button to submit the answer.

The last component included in our **Question** page displays the list of answers using the `map` function on the `Answer` object and `AnswerTile` component to render the answer. Here is the implementation of the `AnswerTile` component:

```
import {
  RocketButton
} from '@/components/ui/atoms/rocketbutton';
import { AnswerBaseType } from '@/graphql/selectors';

export const AnswerTile = (
  { answer }: { answer: AnswerBaseType }
) => {
  return (
    <div className="flex flex-col space-y-2">
      <i className="text-gray-400">
        {`${new Date(answer.createdAt).toLocaleString()}
          ${answer.user.username} answered`}
      </i>
      <div className="text-sm">{answer.content}</div>
      <div className="flex justify-end">
        <RocketButton {...answer} />
      </div>
    </div>
  );
};
```

Inside the `AnswerTile` component, we display the answer content, the date when the answer was posted, the answering user's username, and `RocketButton` to increase the answer's voting score.

Here is the result of our **Questions** page:

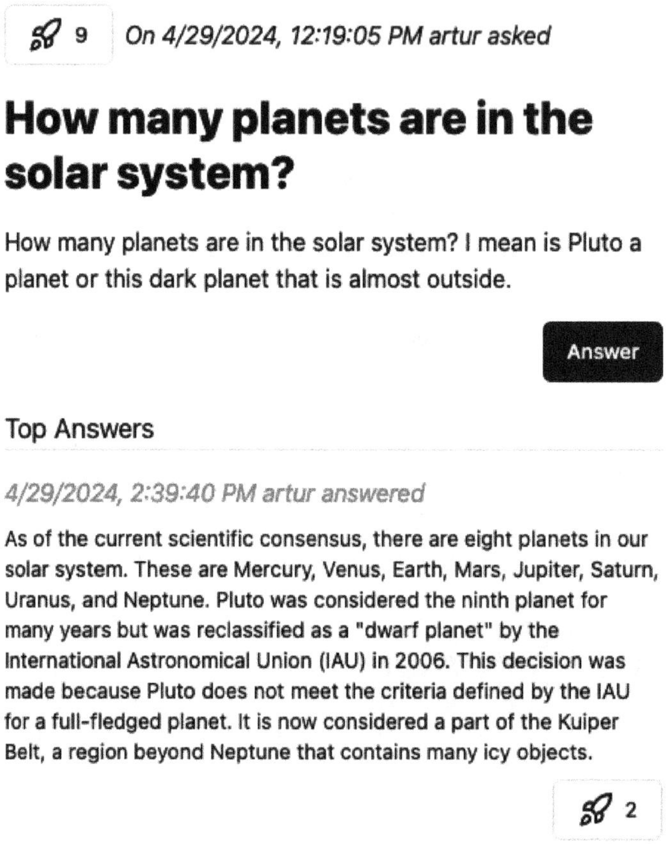

🚀 9   *On 4/29/2024, 12:19:05 PM artur asked*

How many planets are in the solar system?

How many planets are in the solar system? I mean is Pluto a planet or this dark planet that is almost outside.

Answer

Top Answers

4/29/2024, 2:39:40 PM artur answered

As of the current scientific consensus, there are eight planets in our solar system. These are Mercury, Venus, Earth, Mars, Jupiter, Saturn, Uranus, and Neptune. Pluto was considered the ninth planet for many years but was reclassified as a "dwarf planet" by the International Astronomical Union (IAU) in 2006. This decision was made because Pluto does not meet the criteria defined by the IAU for a full-fledged planet. It is now considered a part of the Kuiper Belt, a region beyond Neptune that contains many icy objects.

🚀 2

Figure 9.5: Questions page

That's it – there were lots of components and code, but we have finally finished the frontend!

Summary

In this chapter, we explored how to set up a frontend project. We prepared a GraphQL communication layer that facilitates using the data from the backend inside React visual components. We also learned how to create guard layouts, bootstrapped all components from the ShadCN UI library, and wrote some custom code for Vite and React Router using TypeScript and React.

As you have seen, creating a quality GraphQL-based frontend requires a lot of setup, but it is worth it in the end. That's because later on, we will have less code responsible for communication with the backend inside pages and components.

In the next chapter, we will go back to some theory and explore the security of GraphQL servers.

10

Keeping Data Secure

Crafting a schema-first GraphQL API is undeniably a delightful experience. As developers, we can define our data model and shape our API according to our specific needs. Nevertheless, it is crucial to keep in mind the common vulnerabilities that can arise in any GraphQL implementation.

With GraphQL, clients have the power to request the precise data they need, which is undoubtedly one of its advantages. However, if not properly handled, an obvious vulnerability is the potential to accidentally give a client access to data that they should not have access to. This can lead to inefficiencies or security concerns. For this reason, it is essential to strike a balance between providing flexibility to clients and ensuring that sensitive or unnecessary data is not exposed.

Another vulnerability to consider is the potential for **denial-of-service** (**DoS**) attacks. GraphQL APIs are highly flexible, allowing clients to execute complex and nested queries. While this flexibility empowers clients, it also opens the door to potential abuse. It is crucial to implement rate limiting, query depth limiting, and other safeguards to protect the API server from being overwhelmed by excessive or malicious queries.

Lastly, as with any web application, GraphQL APIs are not immune to injection attacks. It is vital to sanitize and validate user input to prevent malicious queries or mutations if user input is passed to execution environments that need it.

In this chapter, we will look at DoS and injection attacks specifically. By being aware of these common vulnerabilities, we can proactively implement security measures to safeguard our GraphQL APIs. We will also discuss the concept of the query costs calculation, which can protect us from the possibility of feeding the server with too much data.

In this chapter, we will cover the following topics:

- Preventing denial-of-service attacks
- Preventing injection attacks
- Understanding the query cost calculation

Technical requirements

To complete this chapter, you will need a GraphQL IDE such as VSCode, Sublime Text, Notepad++, or GraphQL Editor.

You can access the code in this chapter from the book's GitHub repository: `https://github.com/PacktPublishing/GraphQL-Best-Practices/tree/main/chapter-10`.

Preventing denial-of-service attacks

A DoS attack is a malicious attempt to disrupt or disable the functioning of a network, system, or service. It typically involves overwhelming the targeted resource with an excessive amount of traffic or requests, rendering it unable to handle legitimate user requests. DoS attacks aim to disrupt the availability and performance of a target, often resulting in service disruptions or downtime.

In this section, I will describe GraphQL-specific prevention methods for those attacks.

Depth limiting

Depth limiting in GraphQL refers to the practice of setting a maximum depth for the nested fields that can be queried in a GraphQL request. It is a security measure implemented to prevent potential performance issues and excessive data fetching. Limiting the depth of nested fields ensures that clients cannot query for an excessive amount of data in a single request, thereby optimizing the overall performance and preventing possible abuse of the GraphQL API. This depth limit is typically set by the server and can be configured based on specific requirements and resource constraints.

As GraphQL allows **circular references** – meaning that type A references type B and type B references type A – it is possible for the attacker to construct a query with infinite depth that fetches huge amounts of data at once, freezing our server if we are running on controlled resources. You can see circular references illustrated here:

```
type A {
  b: B
}
type B{
  a: A
}
```

If we don't implement depth limiting on the serverless server, we will receive a huge bill from our serverless provider.

Let's consider a scenario where we have a GraphQL schema with two types: User and Post. The User type has a field called posts that returns a list of Post types, and the Post type has a field called author that returns the User type:

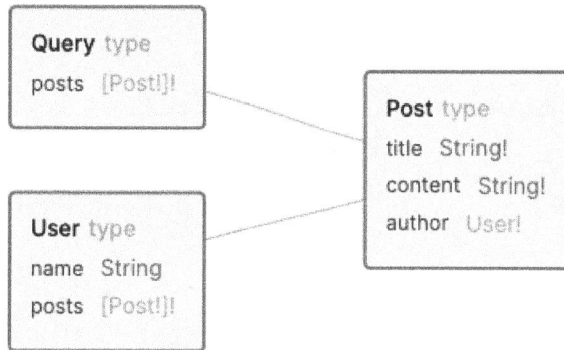

Figure 10.1: Schema with circular references

An attacker could craft a query that recursively requests all the posts for a given user, and for each post, requests the author, creating an infinite loop of requests. This query would result in a substantial amount of data being fetched and potentially overwhelming the server's resources, causing a DoS attack.

Here is an example of a query that can be used for that attack:

```
query Circular {
  posts {
    title
    content
    author {
      name
      posts {
        title
        content
        author {
          name
          posts {
            title
            content
            author {
              name
              # .... and so on
            }
          }
        }
      }
    }
  }
}
```

```
        }
      }
    }
```

To prevent this kind of query, we don't have to change our schema or remove the circular references. Instead, we can limit the depth of our query using our GraphQL server.

Let me present how your algorithm or plugin should work instead:

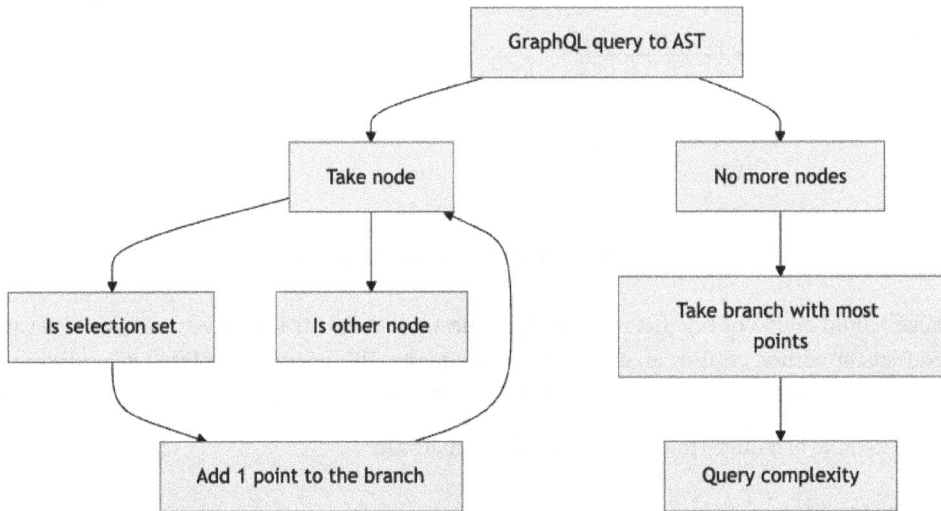

Figure 10.2: Query complexity calculation algorithm

During the parsing of a GraphQL query, the process involves evaluating each field individually. For every field that contains a selection set, we assign a point. The selection set represents nested fields within the query. By assigning points to these selection sets, we can determine the depth of the query. The branch with the highest number of points is considered the query depth. This approach allows us to identify the deepest level of nested fields within the query and establish a clear understanding of its complexity.

So, to calculate query complexity, you need to find the part of the query with the deepest branch. Here is a figure showing how we use the algorithm to limit the queries:

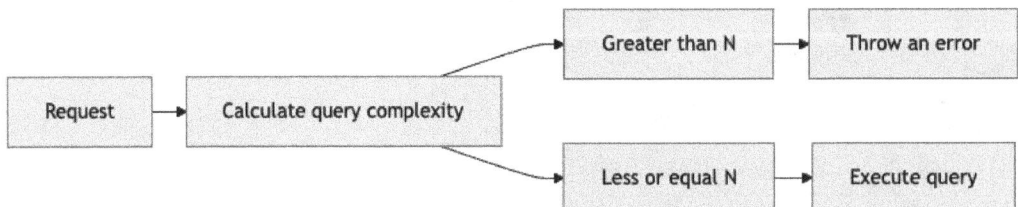

Figure 10.3 – Depth-limiting algorithm

This way, we can know the query complexity and throw an error if it surpasses our depth limit.

Depth limit is not the only danger of a DoS attack in GraphQL. Let's explore another security option: rate limiting.

Rate limiting

Rate limiting is a technique in which a limit is set on the number of requests a client can make to a GraphQL API from a particular IP address and within a particular time frame. By setting a rate limit, whether in GraphQL or REST, developers can ensure that their APIs are not overwhelmed by a single client or a group of clients from the same IP address. This prevents abuse, protects server resources, and helps maintain the overall performance and availability of the API.

This approach is similar to rate limiting in REST APIs, where the goal is also to restrict the number of requests made by a client within a specified time period. Both in GraphQL and REST, rate limiting can be implemented at the server level by tracking the number of requests per IP address and enforcing a predefined limit.

Here is how the rate-limiting algorithm works:

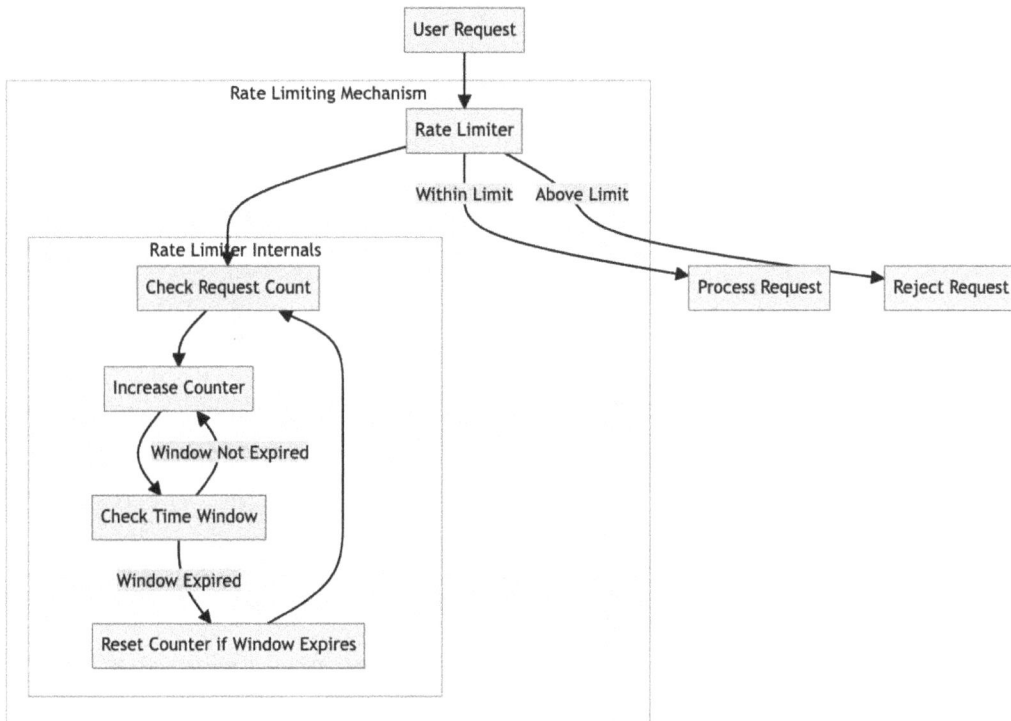

Figure 10.4 – Rate-limiting mechanism

With this kind of algorithm, we can set the limit of requests per time window per user. For example, one user can make 60 requests per minute. If they try to make 61 requests in the minute time window, the system will throw an error.

Implementing request-per-IP rate limiting in GraphQL, as in REST, is an effective way to maintain fairness, protect against abuse, and ensure a reliable experience for all clients interacting with the API. By implementing rate limiting, we can ensure that our service will not unexpectedly become overwhelmed and be unable to handle the demand.

Batching attacks

A **batching attack** refers to the capability of sending multiple queries within a single HTTP request. The level of danger posed by batching attacks is comparable to that of depth-limiting attacks, and what's more, they can be executed without the use of circular references.

Batching attacks are possible because of GraphQL Query Language. GraphQL Query Language supports batching requests – sending many GraphQL requests in one HTTPS call – but an attack can also be made with multiple queries in one network request or field aliases

GraphQL queries can be easily scaled horizontally to simulate a batching attack. Let's take a look at how to do it with aliases:

```
query Q1 {
  p1:posts{
    title
    content
  }
  p2:posts{
    title
    content
  }
  p3:posts{
    title
    content
  }  #... and so on
}
```

Basically, an attacker can create hundreds of thousands of aliases and it will fire hundreds of thousands of resolvers on our GraphQL server.

In order to mitigate the risk of batching attacks utilizing aliases, one preventive measure is to impose a limit on the size of the query – in this case, limiting the number of bytes. This means that the server will not accept the query until it begins parsing it, ensuring that the query does not exceed the size limit. We will do this later, in the *Understanding the query cost calculation* section.

Another situation to consider is when your server supports query batching. In this scenario, an attacker can exploit this capability by sending a malicious request, such as the following:

```
[
  {
    query: < query 0 >,
    variables: < variables for query 0 >,
  },
  {
    query: < query 1 >,
    variables: < variables for query 1 >,
  },
  {
    query: < query n >
    variables: < variables for query n >,
  }
  //...and so on
]
```

This way, they can also send hundreds of thousands of queries at once.

The solution is to limit the number of queries you can send at once. Rather than limiting the number of bites, here we are limiting the number of queries in one request. For example, in GraphQL Yoga, you can set the batching limit inside the server configuration:

```
import { createServer } from 'node:http'
import { createSchema, createYoga } from 'graphql-yoga'

export const yoga = createYoga({
  schema: createSchema({
    typeDefs: /* GraphQL */ `
      type Query {
        hello: String!
      }
    `
  }),
  batching: {
    limit: 2
  }
})

const server = createServer(yoga)
server.listen(4000, () => {
```

```
    console.info('Server is running on
        http://localhost:4000/graphql')
})
```

By setting the batching limit field inside the `createSchema` function, we tell the server to allow only two GraphQL requests at once.

Many GraphQL servers already have configurations to prevent batching attacks. In this section, we analyzed what these attacks are and how to prevent them. In the next section, we will learn how to prevent something unexpected – injections.

Preventing injection attacks

Injection attacks work by exploiting vulnerabilities that arise when user input is not properly validated or sanitized. Attackers can insert malicious code as part of the input, which is then executed within the GraphQL API. This can lead to unauthorized access, data leakage, or even complete compromise of the system.

In this section, we will learn how to prevent NoSQL and SQL injection attacks.

NoSQL GraphQL injection example

NoSQL injections can pose a threat when utilizing filters with custom scalars in GraphQL. A NoSQL injection occurs when an attacker is able to manipulate the filter parameters in a way that exploits vulnerabilities in the underlying NoSQL database. This can lead to unauthorized access, data leakage, or even the execution of arbitrary code within the database.

Consider the following schema that allows users to filter their income data with a custom scalar JSON filter:

```
type User{
  name: String
}

type Income{
  value: Float
  when: String
}

schema{
  query: Query
}

type Query{
```

```
  me: UserQuery
}

scalar JSON

type UserQuery{
  income(
    filter: JSON
  ): [Income!]!
  me: User
}
```

Our backend will need authorization to access the user query and the user's financial income. We also added a JSON filter so we can use the `filter` parameter to set up the MongoDB filter as a schema consumer. What could go wrong?

To filter MongoDB objects inside our backend code, we would enter this code:

```
Db.collection("User").find({
  _id: source._id,
  ...args.filter
})
```

Here, we are just spreading the `filter` argument inside the MongoDB `find` function. In MongoDB, the `find` function is employed to search for and retrieve all objects that meet specific filtering criteria. It allows us to define rules or conditions to match against the data stored in the database. The `find` function scans through the collection and returns all the objects that satisfy the specified filter rules.

Now imagine that an agent provides this code as a parameter of the `filter` argument:

```
filter:{
  _id: {
    $ne: null
  }
}
```

This will return all the income data of all the users inside our MongoDB database, creating a serious security breach. We shouldn't allow this kind of custom JSON scalars for functions used in NoSQL databases; instead, we should provide a specific input for every possible filtering operation.

For example, for this schema, we can provide the following input:

```
type UserQuery{
  income(
    filter: IncomeFilter
```

```
  ): [Income!]!
  me: User
}

input IncomeFilter{
  from: String!
  to: String
}
```

This way, we specify the input format exactly.

Of course, we also need to change the backend code:

```
Db.collection("User").find({
   _id: source._id,
   from: { $gte: args.filter.from },
   ...(args.filter.to ? { to:{ $lte: args.filter.to } }
     : {}
   )
})
```

This makes it impossible to conduct an injection attack.

It is important to keep in mind that custom scalars in GraphQL were not designed to accept arbitrary input without any type and just pass it to the backend. But by properly defining and validating input types for custom scalars, we can mitigate the potential vulnerabilities and ensure the safety and security of our GraphQL backend.

Now we can check how to prevent an injection attack when we have a SQL database connected to our backend GraphQL server.

SQL GraphQL injection example

Now, suppose we have a GraphQL API that allows users to query for a specific product by providing a productId parameter, like so:

```
type Product{
  name: String!
  price: Float!
  _id: String!
}

type Query{
  getProductByID(
    id: String!
```

```
  ): Product
}
```

The API fetches the product details from a SQL database using a raw SQL query.

To demonstrate this behavior, let's look at the example of a vulnerable GraphQL query resolver in Node.js:

```
const getProduct = (productId) => {
  const query = `SELECT * FROM products WHERE id =
    ${productId}`;
  return db.query(query);
}

const resolvers = {
  Query: {
    product: (_, { productId }) => getProduct(productId),
  },
};
```

The productId parameter is directly interpolated into the SQL query without any sanitization or parameterization. This opens up the API to SQL injection attacks. An attacker can exploit this vulnerability by injecting malicious SQL code into the productId parameter.

For instance, consider the following attack payload:

```
query {
  product(productId: "1; DROP TABLE products;") {
    id
    name
    price
  }
}
```

In this example, the attacker appends a semicolon followed by a malicious SQL command (DROP TABLE products) to the productId parameter. If executed, this attack would result in the deletion of the entire products table from the database!

How can we stop this? Here's an example of how to modify the vulnerable code to prevent SQL injection attacks using parameterized queries with a library such as mysql2 in Node.js:

```
const getProduct = (productId) => {
  const query = 'SELECT * FROM products WHERE id = ?';
  return db.query(query, [productId]);
}

const resolvers = {
```

```
  Query: {
    product: (_, { productId }) => getProduct(productId),
  },
};
```

By using parameterized queries, the `productId` parameter is treated as a value rather than being directly interpolated into the SQL query. This effectively neutralizes SQL injection attempts and ensures the security of the GraphQL API in Node.js by telling SQL that we are inserting an ID here, not a SQL instruction.

Preventing injection attacks within a GraphQL API is crucial to maintaining the security and integrity of the system. That is why we need to know how to prevent injection and about rate limit and depth limit. But on top of that, those limitation methods can be combined into one powerful query cost calculation algorithm, which we will look at next.

Understanding the query cost calculation

In DoS attacks, we have many different possibilities of feeding the server with too much data, and we can protect ourselves from all of them by using the query cost calculation method. This requires additional middleware and extra work from us, but it should be worth implementing on bigger APIs.

Query cost calculation in GraphQL involves assigning a numerical value to each field in the schema to determine the overall cost of executing a given query. This allows for resource management and optimization, ensuring that queries do not consume excessive server resources.

Let's explore another example schema to understand how query cost calculation works. In this schema, we will have `Product`, which can have `Reviews`, and each `Reviews` has a corresponding `Author`. Here, we'll assign a `cost` value to each field based on its complexity and the resources it requires. For simplicity, we'll use a scale from 1 to 10, where 1 represents a low-cost field and 10 represents a high-cost field. We'll also assume that any field without the cost directive is worth 1 point.

Here is the schema:

```
type Product{
  id: ID!
  name: String!
  price: Float!
  description: String @cost(value: 4)
  reviews: [Review]
}

type Review{
  id: ID!
  rating: Int!
  comment: String! @cost(value: 2)
```

```
    author: User!
}

type User{
  id: ID!
  name: String!
  email: String
}

type Query{
  productById(
     id: ID!
  ): Product
}

schema{
  query: Query
}

directive @cost(
  value: Int
) on FIELD_DEFINITION
```

In this scenario, we have assigned cost values to the comment and description fields. This is because these fields likely contain long strings, which can be resource-intensive in terms of bandwidth and server execution.

We know that description can be very long and take the most space, which is why I assigned it 4. Meanwhile, the comment content should be shorter (though it's still a bigger field), which is why I assigned it 2. By assigning cost values to these fields, we can better manage and optimize resource allocation.

Now, let's look at the following query and then calculate its cost:

```
query Cost($productById_id: ID!) {
  productById(id: $productById_id) {
    id
    name
    price
    description
    reviews {
      id
      rating
      comment
      author {
```

```
            name
            email
        }
      }
    }
  }
```

This query allows us to retrieve a single product by its unique identifier. Along with the product information, it also fetches the associated reviews and the respective authors of those reviews.

The total cost of this query is as follows:

```
1+1+1+4+1+1+2+1+1 = 13
```

In our implementation, we have included every scalar field in the query. For each scalar field, using the @cost directive, we multiplied its cost by the cost multiplier if it was specified in the original schema. If the cost multiplier was not set for a field, we applied a default multiplier of 1. By doing that, we know that our query cost is 13.

Now we can add another field to our cost directive. Take a look at the following multiply argument inside the cost directive:

```
type Query{
   productById(
      id: ID!
   ): Product
   products(
      limit: Int!
   ): [Product!]! @cost(value: 1 multiply: "limit")
}
directive @cost(
   value: Int!
   multiply: String
) on FIELD_DEFINITION
```

Here, we have introduced a new feature to enhance the functionality of the @cost directive. This new feature allows users to multiply specific field arguments by a given value factor. This addition aims to provide more flexibility and control to our users when calculating costs.

Now you can try to calculate the cost of the following query based on the previous schema code:

```
query CostMulti{
   products(limit: 20) {
      name
      price
      description
```

```
    }
}
```

The total cost of this query is as follows:

```
20*1*(1+1+4) = 120
```

To calculate this query cost, we multiplied every product query cost by the `limit` value inputted by the schema consumer, which is `20`, and the `value`, specified by the `@cost` directive, which is `1`. Now we can easily control incoming queries by setting a global limit on our GraphQL server.

Overall, the implementation of query cost analysis serves as a protective measure for our GraphQL server against potential attacks involving deep and horizontally large queries. In addition, it is considered a best practice to discourage schema consumers, specifically developers, from writing excessively large queries. By implementing query cost analysis, we can mitigate the risks associated with these types of queries and ensure the optimal performance and security of our GraphQL server.

Summary

In this chapter, we have learned about some of the possible attacks on GraphQL servers. In our analysis, we looked at DoS attacks in GraphQL and how to prevent them, including establishing a depth limit for queries and implementing a rate limit for our server. Furthermore, we delved into the mechanisms behind batching attacks and the potential risks associated with injection attacks.

To further enhance our understanding, we also covered the concept of calculating the cost of a query using GraphQL directives. By comprehensively studying these various aspects, we aim to strengthen the security measures and overall resilience of our GraphQL implementation.

In the next chapter, we will look at how to handle GraphQL errors properly.

11

Describing Errors in GraphQL

As developers, we know that errors are an inevitable part of building any application. Luckily, GraphQL provides us with powerful tools to handle them gracefully. By defining errors using GraphQL types and following established conventions, we can enhance the reliability and schema consumer experience of our GraphQL APIs.

In this chapter, we will explore the essential concepts and best practices for effectively describing errors in GraphQL. We will start by understanding how we throw errors, before learning how to specify errors with enums, and fully describe them with union-based responses. By the end of the chapter, we will learn why we need to mask all errors eventually and whitelist only those wanted by the developer on production.

In this chapter, we will cover the following topics:

- Understanding how we throw errors in GraphQL-based servers
- Using enums for error codes
- Treating errors as data
- Treating responses as unions
- Understanding error masking

Technical requirements

To complete this chapter, you will need a GraphQL IDE such as VSCode, Sublime Text, Notepad++, or GraphQL Editor.

You can access the code in this chapter on the book's GitHub repository at `https://github.com/PacktPublishing/GraphQL-Best-Practices/tree/main/chapter-11`.

Understanding how should we throw errors in GraphQL-based servers

Understanding how GraphQL throws errors is crucial for developers working with GraphQL. When an operation encounters an error, the GraphQL server throws an error response with detailed information about what went wrong. By familiarizing ourselves with the error-handling mechanism, we can effectively troubleshoot and improve the overall reliability of our GraphQL applications.

Let's consider a simplified GraphQL schema that includes a `Query` type. Within this `Query` type, there is a field called `withoutErrors`. Take a look at the code of the schema:

```
type Post{
  title: String
  content: String
}
type PostResponse{
  posts: [Post!]!
}
type Query{
  withoutErrors: PostResponse
}
```

This schema should just returns `PostResponse` on the `withoutErrors` field.

Next, let's code the `resolver` function for the `withoutErrors` field query. Here's an example of `resolver` code that throws an error:

```
const resolvers = {
  Query: {
    withoutErrors: () => {
      throw new Error("Standard GraphQL Error");
    },
  },
};
```

Then we can write a query to trigger the `resolver` code:

```
query WithoutErrors {
  withoutErrors {
    posts {
      title
      content
    }
  }
}
```

When an error is thrown, GraphQL will include it in the response sent back to the client. Along with the error message, the response will also include relevant error information, such as the path of the field where the error occurred and any error extensions you may have attached.

Now, in the following example, I will show you the built-in error mechanism. You will see how the default response looks if the `withoutErrors` query encounters an error:

```
{
  "errors": [
    {
      "message": "Standard GraphQL Error",
      "locations": [
        {
          "line": 2,
          "column": 3
        }
      ],
      "path": ["withoutErrors"]
    }
  ],
  "data": null
}
```

By leveraging this built-in error-handling mechanism, GraphQL allows clients to receive both partial data and error information in a single response. This enables developers to handle errors gracefully and provide meaningful feedback to the client. The flexibility of GraphQL's error-handling mechanism allows you to build robust and reliable applications, but it might not be enough for big production environments.

In the next section, we will start adding details to our errors using simple enums to provide more data about the error to the schema consumer.

Using enums for error codes

It is important to be mindful of the granularity of errors. While it may be tempting to return a single generic error for all scenarios, it is better to provide more specific error messages for individual scenarios. This allows clients to understand the root cause of the error and take appropriate actions. By being specific, we can guide developers toward the right solutions and save valuable debugging time.

By specifying an enum with error variants, we can return errors as a valid GraphQL response. Then the schema consumer – depending on which `ErrorVariant` is returned – can handle it in their own way.

In this short example, we will specify an enum to be used as an error descriptor. Here is the schema:

```
type Post{
  title: String
  content: String
}
enum ErrorVariant{
  NOT_FOUND
  FORBIDDEN
}
type PostResponseVariant{
  posts: [Post!]
  errors: [ErrorVariant!]
}
type Query{
  enumErrors: PostResponseVariant
}
```

In this schema, our response can contain the posts and errors fields. Plus, errors are described as enum variants, so when the error our backend throws is expected, we can see it as an enum field.

We also need a GraphQL resolver that does not throw an error, but instead returns an enum for the expected error, informing the user about the cause of the error. Here is the resolver code:

```
const resolvers = {
  Query: {
    enumErrors: () => {
      return {
        errors: ["FORBIDDEN"]
      }
    }
  },
  PostResponseVariant: {
    posts: () => [{ title: "Hello", content: "World" }]
  }
};
```

Next, we will execute the query on our schema. Take a look at the following query:

```
query ErrorVariant {
  enumErrors {
    posts {
      title
      content
    }
```

```
        errors
    }
}
```

With this query, we want to fetch the list of posts and see the errors as data. The query should return a response with an error that is an `ErrorVariant` enum field.

In the `resolver` implementation, we throw an error on every request. The error response for the query would look like this:

```
{
"data":{
  "enumErrors": {
    "errors": [
      "FORBIDDEN"
    ]
  }
}
}
```

This is so that we can easily predict when this error happens and prepare a message for the end user based on this enum field, such as **Sorry, you can't access this part of our service**.

Looking at this method of using enums, there are a number of advantages:

- It is easily translatable, meaning you can use your own error messages on the frontend
- It is API friendly, meaning you can easily detect what error was thrown when using your API in other backend solutions
- It can throw multiple errors in an array
- Schema consumers can provide their own custom way of handling errors, which allows for more precision of error descriptions.

However, there are also some disadvantages:

- Usually, we need to attach documentation to every enum field error code, and the schema consumer needs to read them all to understand them
- They can contain both `posts` and `errors` types, which means we can't always be sure whether the response is correct based on the response type
- The schema consumer needs to provide their own way of handling every error – I already mentioned this as an advantage, but the disadvantage is that it requires extra time to provide a mechanism for handling every enum field

So, despite some disadvantages, when it comes to implementing and obtaining the most value from returning errors, utilizing enum fields is considered the optimal choice. This approach offers advantages in terms of implementation efficiency and overall value.

In the following section, we will delve into specific techniques for enhancing the precision and specificity of error returns.

Treating errors as data

Another possible method to describe errors in GraphQL is to treat them as data. Unlike traditional REST APIs, where errors are often returned as HTTP status codes, GraphQL encourages us to model errors using GraphQL types. This approach enables clients to understand and describe errors in a more structured and predictable manner. This way, schema consumers can receive fully described expected error descriptions and simply display them to the end user.

Let's take a look at an example. In the following schema, we created a dedicated PostError type to keep the structure of the error response inside our GraphQL schema:

```
type Post{
  title: String
  content: String
}
enum ErrorVariant{
  NOT_FOUND
  FORBIDDEN
}
type PostError{
  message: String!
  variant: ErrorVariant
}
type PostDescribedResponse{
  posts: [Post!]
  errors: [PostError!]
}
type Query{
  describedErrors: PostDescribedResponse
}
```

The PostDescribedResponse type returned from Query.describedErrors contains two fields – posts and errors. The posts section of the response simply contains an array of posts, while the errors section provides an array that describes any errors encountered with the posts. Each error has its own enum error code and message field with a further description of the error.

Now we will write the query to fetch data from our schema:

```
query ErrorDescribed {
  describedErrors {
    posts {
      title
      content
    }
    errors {
      message
      variant
    }
  }
}
```

Within this query, we want to get a list of posts, and if the query is not successful, we want to receive structured errors with `message` and `variant` fields.

Then the `resolver` code will always return the structured error to show you how those errors should be specified on the backend. This is the `resolver` code:

```
const resolvers = {
  Query: {
    describedErrors: () => {
      return {
        errors: [{
          variant:"FORBIDDEN",
          message:"Ask your team administrator to give you
            access to post resource"
        }]
      }
    }
  },
};
```

Here, we are not only returning the `ErrorVariant` enum status but also receiving an additional message describing what the reason for the error is. While this approach may not be considered API-friendly in terms of code consumption, it can be beneficial for frontend applications. This is because frontend applications have the capability to directly display error messages to end users. Therefore, using this method allows for a more seamless communication of errors to the end user.

Here is the error response example:

```
{
  "data": {
    "describedErrors": {
      "errors": [
        {
          "variant": "FORBIDDEN",
          "message": "Ask your team administrator to give
            you access to post resources"
        }
      ]
    }
  }
}
```

Here, the error message and variant are included in one response. The error is described by an enum `variant` field and the dedicated `message` field. This way, we give the schema consumer a choice to display the message to the end user or provide their own message using the enum field.

Looking at this method, here are the advantages:

- You can provide very descriptive errors
- You can display an error message for the final user of the system by passing the same message received from the backend to a frontend notification, for example

Here are the disadvantages:

- It is not easily translatable – we would need to add another mechanism to provide a locale to the server to make the message localized
- Responses can contain both the `posts` and `errors` fields, so we cannot determine whether the response status returns without errors just from the response type

So, through this method, we have the ability to incorporate specific messages into the errors returned from the backend in GraphQL. In the next section, we will take it a step further by providing responses as unions, which allows us to determine whether the response type corresponds to an error or not.

Treating responses as unions

In addition to the previous methods, another approach is to treat our response as a union. Unions in GraphQL provide us with the ability to return one of multiple possible types. This is particularly advantageous because it enables us to easily identify whether the response contains only the `errors` field or only the `posts` field.

Additionally, GraphQL servers include the __typename property on every returned object, which allows the consumer of the schema to accurately determine the type of the returned object.

In the following schema, we will introduce a union type that will be able to represent both successful responses and responses that indicate errors:

```
type PostResponse{
  posts: [Post!]!
}

enum ErrorVariant{
  NOT_FOUND
  FORBIDDEN
}

type PostError{
  message: String!
  variant: ErrorVariant
}
type ErrorResponse{
  errors: [PostError!]!
}
union ResponseUnion = PostResponse | ErrorResponse
type Query{
  unionErrors: ResponseUnion
}
```

The ResponseUnion type is a union type that encompasses two possible types of responses:

- The first is PostResponse, which indicates a successful response and includes a list of posts

- The second is ErrorResponse, which indicates an expected error

This distinction allows us to handle both successful and error responses appropriately.

Now let's write the resolver code that will randomly return ErrorResponse and PostResponse:

```
const resolvers = {
  Query: {
    describedErrors: () => {
      if(Math.random() > 0.5){
        return {
          errors: [{
            variant:"FORBIDDEN",
            message:"Ask your team administrator to give
              you access to post resource"
```

```
            }],
          }
        }
      return {
        posts:[{ title: "Hello World", content: "This is
          the message" }]
      }
    }
  },
};
```

Here, we return the errors and correct responses randomly inside the `resolver` code. We also return an error with an enum field and a message.

Having response union types allows the schema consumer to determine the response type when writing the query and provides separate mechanisms for both response types. Here is the query:

```
query UnionErrors {
  unionErrors {
    ... on PostResponse {
      posts {
        title
        content
      }
      __typename
    }
    ... on ErrorResponse {
      errors {
        message
        variant
      }
      __typename
    }
  }
}
```

Inside this query, we differentiate the response using inline fragments on the `PostResponse` and `ErrorResponse` types. Then, by requesting for the `__typename` field, we can be completely sure which kind of response was returned inside our codebase.

After executing our `UnionErrors` query, we should either get an error or a valid response. Here is the data returned for the error response:

```
{
  "data": {
    "unionErrors": {
      "__typename":"ErrorResponse",
      "errors": [
        {
          "variant": "FORBIDDEN",
          "message": "Ask your team administrator to give
            you access to post resource"
        }
      ]
    }
  }
}
```

We can see that the returned response contains the __typename field which value isErrorResponse. It consists of the error variant and the error message.

Our resolver also returns the array of posts. Let's take a look at the response:

```
{
  "data": {
    "unionErrors": {
      "__typename":"PostResponse",
      "posts":[{"title":"Hello", "content":"World"}]
    }
  }
}
```

As a schema consumer, we can differentiate the correct response from error one using the following code:

```
if(response.data.unionErrors?.__typename === "ErrorResponse"){
  alert("Error")
}
if(response.data.unionErrors?.__typename === "PostResponse"){
  alert("Correct")
}
```

Using this kind of code on GraphQL Server responses allows us to differentiate response type by the __typename field.

In conclusion, it is evident that using union error types provides the most descriptive approach when returning errors as a response. By utilizing union types, we can clearly distinguish between different types of errors and handle them accordingly. This enhances the clarity and effectiveness of error handling in our response structure.

In the next section, we will try to hide errors that shouldn't be seen by schema consumers.

Understanding error masking

Error masking is a GraphQL technique that allows you to control the level of detail exposed in error messages sent to clients. It helps to prevent sensitive information from being leaked in error responses and promotes better security and privacy practices.

Let's explore how error masking works with an example. First, start by bootstrapping an `axolotl` project inside your command line:

```
npx @aexol/axolotl create-yoga
```

Then replace the `./schema.graphql` file with the following content:

```
type Query{
  error: String!
  errorMasked: String!
}

schema{
  query: Query
}
```

Here, we have two resolvers to show you a normal error and a masked error. Ideally, your server framework should handle error masking under the hood, but I will show you how to create your own masked error mechanism.

Go to `src/index.ts` and change the file's content to turn off default error masking in GraphQL Yoga by setting the `maskedErrors` property to `false`:

```
import { graphqlYogaAdapter } from '@aexol/axolotl-graphql-yoga';
import resolvers from '@/src/resolvers.js';

graphqlYogaAdapter(resolvers, {
  yoga: {
    maskedErrors: false,
  },
}).listen(parseInt(process.env.PORT || '4000'), () => {
  console.log('LISTENING to ' + process.env.PORT ||
```

```
      '4000');
});
```

Now we will write the resolver code:

```
import { createResolvers } from '@/src/axolotl.js';

class MaskedError extends Error {
  constructor(message: string, unmask?: boolean) {
    super(!unmask ? 'Unexpected error' : message);
  }
}

const resolvers = createResolvers({
  Query: {
    error: () => {
      throw new MaskedError('Invalid resolver coded',
        true);
    },
    errorMasked: () => {
      throw new MaskedError('Invalid resolver coded');
    },
  },
});

export default resolvers;
```

We started by creating our own error class that extends the original error. Within this class, we require an unmask? Boolean property to display (or not display) the error to the schema consumer.

With that, inside the command line at the root of the project, run the following command to start the GraphQL server for this project:

npm run dev

GraphQL should be running on localhost:4000/graphql.

Then, after executing the following query:

```
{
  error
}
```

We should receive the following response:

```
{
  "errors": [
    {
      "message": "Invalid resolver coded",
      "locations": [
        {
          "line": 2,
          "column": 3
        }
      ],
      "path": [
        "error"
      ]
    }
  ],
  "data": null
}
```

We can see the original error message here as the error was not masked. However, we will now execute the second query:

```
{
  "errors": [
    {
      "message": "Unexpected error",
      "locations": [
        {
          "line": 2,
          "column": 3
        }
      ],
      "path": [
        "errorMasked"
      ]
    }
  ],
  "data": null
}
```

In addition to that, our `resolver` code for `errorMasked` fields throws the same error. However, we have implemented an error masking mechanism to conceal the original error message. This approach ensures that the error details are hidden from the end user, providing an added layer of security and privacy.

It is important to note that we do not always want the schema consumer to know the reason for the unhandled exception, so the best practice is to mask all the unhandled errors in production mode. Not masking errors could lead to unveiling sensitive information, such as the following:

- Authentication errors specifying whether the username or password is incorrect can lead to brute-force attacks on the system.

- Internal server errors and unexpected errors can accidentally expose some confidential data and information about the underlying technology stack. This knowledge can aid attackers in identifying potential vulnerabilities or weaknesses to exploit.

- Configuration errors such as misconfigured connection strings or API keys may expose environment variables in the error message.

By implementing error masking techniques, you can ensure that sensitive fields are not exposed in error responses, enhancing the security and privacy of your GraphQL API.

Summary

As in every backend system, handling errors properly is important inside GraphQL-driven servers.

In this chapter, we learned how GraphQL handles errors and how, step by step, we can provide better methods for returning errors. We started by implementing simple enums for error variants and ended with structured error types with union responses.

In the next chapter, we will delve into the art of writing comprehensive documentation both within and beyond the schema.

Documenting your Schema

In this chapter, we will explore the topic of writing documentation for your GraphQL schema. The documentation should be a signpost for both the schema developer and schema consumer, helping the schema to be created and understood and to avoid redundant questions about it.

We have already explored how documentation strings work in GraphQL, but in this chapter, we will try to describe the nodes from the Questions and Answers schema from *Chapter 8*. By doing so, we will cover the differences between documenting types, interfaces, and input GraphQL nodes.

Also, we will learn how to write additional Markdown documentation, as well as try to guide documentation readers by providing them with root operation descriptions.

In this chapter, we will cover the following topics

- Writing documentation for GraphQL interfaces and types
- Crafting Markdown documentation for root queries and mutations.

Technical requirements

To complete this chapter, you will need a GraphQL IDE such as VSCode, Sublime Text, Notepad++, or GraphQL Editor.

You can access the code in this chapter from the book's GitHub repository: `https://github.com/PacktPublishing/GraphQL-Best-Practices/tree/main/chapter-12`.

Writing documentation for GraphQL interfaces and types

To begin, let's start by explaining the purpose of our GraphQL schema.

The main purpose of a schema is for the schema consumer, or someone who will implement the schema, to understand a system's functionalities without having to ask the schema creator additional questions. A well-documented schema can be compared to Plato's cave allegory, where the idea is represented by the illuminated objects within the cave, while the actual object itself is represented by the shadows cast on the cave walls.

In this analogy, the documented schema serves as a representation of the underlying system, providing developers with a clear understanding of its structure, relationships, and capabilities. Just as the illuminated objects in the cave provide a conceptual understanding of the world outside, the documented schema provides developers with an abstract representation of the system.

Now we will use the schema created in *Chapter 8* and try to document all the types and fields from that schema. Since the majority of types within the schema outlined in *Chapter 8* are based on GraphQL interfaces, it is essential to prioritize documenting these interfaces.

Interfaces

First, we will document how objects in our service are distinguished and what makes them unique across the service. Here is the documentation of the `StringId` interface used to mark the database objects:

```
"""
An interface representing an object with a string ID.
"""
interface StringId {
  """
  The string ID of the object uses a unique identifier in
    MongoDB.
  https://www.mongodb.com/docs/manual/reference/method/
    ObjectId/
  """
  _id: String!
}
```

The `StringId` interface is responsible for holding the hexadecimal string representation of MongoDB's `ObjectId`. In this documentation, we explain this concept and then provide a documentation link so the schema reader can learn more about the object, including what data can be extracted from it.

Then, we need to describe the interface responsible for the object created and the modified dates:

```
"""
An interface representing an object with the created and updated dates
and times. The format is the ISO string format:
yyyy-MM-dd'T'HH:mm:ss. SSSXXX
for example:
2000-10-31T01:30:00.000-05:00
https://en.wikipedia.org/wiki/ISO_8601
"""
interface Dated {
  createdAt: String!
  updatedAt: String!
}
```

As we marked them as the `String` type, it is important to provide the format that is accepted by the service and an example of how we can construct such a string. In this example, we stated that the ISO format should be used and provided a link to format documentation the schema user can use for extra help.

The `Owner` interface – used to mark types and interface ownership – also needs additional documentation:

```
"""
Every type with this interface implemented has its owner in the
database. Only the owner of the object can modify it using mutations.
"""
interface Owned{
  user: User!
}
```

Here, the documentation states that the `Owned` interface adds a `user` field to each type or interface to notify the schema consumer about the owner of the object. It also states that those types of objects implementing the `Owned` interface can only be modified/created by owners.

The last interface – `Message` – is responsible for controlling the shared fields of the `Question` and `Answer` types. Here is how we will document this:

```
"""
An interface representing a message object. Can be both Question and
Answer.
"""
interface Message implements StringId & Dated & Owned{
  """
  The content of the message.
  """
  content: String!
  """
  The score of the message. Messages can be only upvoted.
  One user can give an infinite number of votes per
  message.
  """
  score: Int!
  """
  The answers to the message.
  """
  answers: [Answer!]!
  _id: String!
  createdAt: String!
  updatedAt: String!
  """
```

```
    Author of the post.
    """
    user: User!
}
```

Here, we documented that `Question` and `Answer` types implement the `Message` interface. We also documented the mechanics of the `Message` score field and explained that every message can be answered.

With that, we have defined the interfaces. We started with interfaces because they define the subsequent objects in our schema. However, there is another important issue related to this. You may have noticed that even though interfaces inherit fields, in GraphQL, you still have to write the fields again, which goes against the principle of "Don't Repeat Yourself." It may seem senseless to rewrite these fields for each implementing `interface` and `type`, but it has its advantages:

- The field of an interface may have different purposes for different objects. Allowing the ability to write separate documentation for each type or interface gives you the possibility to better describe the fields.

- A GraphQL schema is intended for reading, not writing. This means that when using a schema, it is not necessary for the consumer to search for interfaces that a type is implementing. Instead, the consumer can easily access the relevant documentation directly.

So, now that we have documented the interfaces, let's document the types that implement those interfaces.

Types

There are two types to look at in this section – the `Question` and `Answer` types.

We need to document them, even though they inherit most of their fields from the interfaces, as we don't want the schema consumer wasting time searching for the interfaces. Moreover, descriptions of fields may differ across the types implementing the same interface, so there's an extra need for documentation.

Let's start with the `Question` type:

```
"""
A type representing a main object in our service. It contains
questions asked by users. User needs to be logged in to ask a
question.
"""
type Question implements Message & StringId & Dated & Owned{
    """
    The content of the question.
    """
    content: String!
    """
```

```
    The score of the question. Messages can be only upvoted.
    One user can give an infinite number of votes per
    message.
    """
    score: Int!
    """
    The string ID of the question.
    """
    _id: String!
    """
    The answers to the question.
    """
    answers: [Answer!]!
    """
    The title of the question. The title will be displayed in
    the search results and should contain a shortened version
    of a question.
    """
    title: String!
    """
    The date and time when the question was created.
    """
    createdAt: String!
    """
    The date and time when the question was last updated.
    """
    updatedAt: String!
    """
    The user who wrote the question.
    """
    user: User!
}
```

All of the fields already exist in the interfaces previously described, except for the `title` field. Nevertheless, we need to document every `type` field to make them accessible for schema consumers and tools such as code generators.

Moving on, inside the `Answer` type, we will also have all the fields from the `Message` interface, as well as a `to` field:

```
"""
Represents object that is an answer to a question or is an answer to
another answer.
"""
type Answer implements Message & StringId & Dated & Owned{
```

```
    """
    Represents an object which this answer is answering.
    """
    to: ToAnswer
    """
    The content of the answer.
    """
    content: String!
    """
    The score of the answer. Messages can be only upvoted.
    One user can give an infinite number of votes per
    message.
    """
    score: Int!
    """
    The string ID of the answer.
    """
    _id: String!
    """
    The answers to the answer.
    """
    answers: [Answer!]!
    """
    The date and time when the answer was created.
    """
    createdAt: String!
    """
    The date and time when the answer was last updated.
    """
    updatedAt: String!
    """
    The user who wrote the answer.
    """
    user: User!
}
```

Here, we only documented the Answer type and the to field, which represents the answer or question being answered. After the changes, we have updated the documentation for the fields coming from the Message interface to now refer to an answer instead of a message. This is important as the role of the fields has changed – both the Answer and Question types implement the Message interface but the role of the fields is now different.

At this point, we have written the documentation for the main interfaces and types. Now, we want to write the documentation for our input types as well.

Input types

Creating documentation for input types is important as we want to tell our schema consumers what kind of values our GraphQL server expects. This is useful in many situations, especially the following:

- There is a character limit on a string. In this case, we can provide that information inside the input field documentation.

- There is some additional validation on the field, for example, an email address or phone number. Here, we should provide information inside the documentation so that the schema consumer can pass the right value.

With that, we will start by documenting the `CreateQuestion` input type:

```
"""
The input for creating a question. The user needs to be logged in to
create a question.
"""
input CreateQuestion{
    """
    The longer version of the question should contain no more
    than 2000 characters.
    """
    content: String!
    """
    The title of the question should be 255 characters
    maximum for better readability.
    """
    title: String!
}
```

Here, we describe the `CreateQuestion` input and remind the schema user about the need for authorization. Then, we describe good practices for the `content` and `title` fields.

Next, we will document the `CreateAnswer` input type:

```
"""
The input for creating a question. The user needs to be logged in to
create an answer.
"""
input CreateAnswer{
    """
    The content of the answer should contain no more than
    2000 characters.
    """
    content: String!
```

```
"""
The id of the Question or Answer which this answer
answers.
"""
to: String!
}
```

Specifically, documenting the `to` field is really important as, due to the field name, it would be hard to guess what that field refers to without the documentation.

Now, to sum up what we've done so far, let's list the documentation best practices:

- **Provide a clear description**: Begin by providing a concise and informative description of the type. Explain its purpose, its intended use, and any important details that developers need to know.

- **Cross-reference implementing types**: If there are specific types that implement the interface, provide links or references to their documentation. This allows developers to easily navigate between the interface documentation and the implementing types.

- **Describe input fields' limits and validation**: If the input field type is `String`, specify what kind of string it is, such as an email or some custom date format. Giving additional documentation makes the context more understandable.

- **Update the documentation**: As the schema evolves and changes over time, make sure to keep the documentation up to date. When modifications are made, ensure that the documentation reflects these changes accurately to avoid confusion or inconsistencies.

As we have learned how to document types, interfaces, and input types, we are now ready to write better documentation for the GraphQL operations with Markdown syntax.

Crafting Markdown documentation for root queries and mutations

For the types and inputs, it is sufficient to provide short textual documentation. However, for Markdown documentation, we should include more detailed information for the schema consumer. This way, they can easily select the right operation for the desired purpose. While you may think it can bloat the schema, remember the schema is mainly for the schema consumer and their understanding.

We will begin by describing our root `Query` type:

```
type Query{
  """
  Search for questions based on a query string. It will
  search through the question titles and respond with an
  array of questions paired with the most voted answer. You
```

can even search for a topic, like this:

```gql
query Search{
  search(query: "space"){
    question{
      _id
      title
      createdAt
      user{
        username
      }
    }
    bestAnswer{
      content
      user{
        username
      }
    }
  }
}
```

This way you should use the search query.
"""
search(
"""
The query string to search for questions.
"""
query: String!
): [QuestionsResponse!]!
"""
This operation retrieves the ten highest voted questions.
It will return an array of questions along with their
corresponding most voted answer.
"""
top: [QuestionsResponse!]!
"""
Gets a question by its ID.
"""
question(
  """
  The ID of the question.

```
    """
    _id: String!
  ): Question
  """
  Gets the currently authenticated user data.
  """
  me: User!
}
```

We started by documenting the search operation, and as GraphQL documentation supports Markdown, we also added a GraphQL Query Language example to it. The query states how to use this operation and what fields we should request from the returned type to display the search results.

Then, in the top field, we have specified that it should return the top 10 questions with the highest number of votes. The response type for this field will be the same as when using the search operation.

The question field of the root query fetches one question by its ID and the user field returns the authorized user.

Additionally, it is specified that we should include the user field in the request to display the author of the question. Similarly, we need to request the bestAnswer author. Furthermore, we should also request the createdAt field to indicate the age of the question.

After documenting all the root queries, we can document our root mutations. In the root Mutation type, we have two pipes we have to document, like so:

```
type Mutation{
  """
  Use pipes to authorize user actions. You need to pass
  Authorization headers to resolve the fields of
  Mutation.user, for example:
  ```json
    {
      "headers": {
      "Authorization": "YOUR_JWT_TOKEN"
      }
    }
  ```

  To obtain the token, you must execute the login
  operation from the Mutation.public
  PublicMutation.login. Then you can then pass this token
  as an authorization header to all mutations beginning
  from Mutation.user
  """
```

```
    user: UserMutation
    """
    Use this mutation for all operations that don't need
    user authorization.
    """
    public: PublicMutation
}
```

We began by defining the `Mutation.user` field and specified that when resolving the fields of `UserMutation` returned from this field, a JWT must be provided. This token can be obtained from the login mutation.

Furthermore, we mentioned that the `publicMutation` and `userMutation` fields serve as pipes to other mutations that are the same field types.

It is important to note that the `public` field does not require an additional `Authorization` header, so all its `resolvers` can be accessed by the schema consumer without the authorization mechanism

We will describe those fields inside `PublicMutation` now:

```
"""
Mutations for authentication. Use without authorization.
"""
type PublicMutation{
  """
  Register using a username and password. Passwords are
  stored inside database using salt & hash mechanism.
  AuthPayload contains the `token` that needs to be used in
  authorization process and the User object.
  """
  register(
    username: String!
    password: String!
  ): AuthPayload!
  """
  Log in using a username and password. Passwords are
  stored inside database using salt & hash mechanism.
  AuthPayload contains the `token` that needs to be used in
  the authorization process and the User object.
  """
  login(
    username: String!
    password: String!
  ): AuthPayload!
}
```

We started by describing that `PublicMutation` field mutations will be used for authentication purposes. Then, we documented the `login` and `register` fields. We also described how they should work on the backend side.

Next, we will document the `UserMutation` type to get all the authorized operations described:

```
"""
Those mutations can be run only by a logged in user. They need the
Authorization header with the token value present in the request.
"""
type UserMutation{
  """
  This field should be used to post a question. Here is the
  example GQL query:

  ```gql
  mutation PostQuestion{
    user {
      postQuestion(createQuestion: {
        title:"Are we alone in space?",
          content:"I wonder what happens when we find out
            another species"
      })
    }
  }
  ```
  """
  postQuestion(
    createQuestion: CreateQuestion!
  ): String
  """
  Post answer to a question or answer. To do so, provide
  the question or answer _id.
  """
  postAnswer(
    createAnswer: CreateAnswer!
  ): String
  """
  Vote for a question or answer by providing its id. Every
  question can be voted on an infinite number of times by
  every user.
  """
  vote(
```

```
    _id: String!
  ): Int
}
```

In the `UserMutation` description, we reminded the schema consumer that they should pass the `Authorization` header with the JWT value to be able to access its fields. The first of those mutation fields, the `postQuestion` field, gave us a clear example of how to use `UserMutation`. `postQuestion` together with the query example.

By incorporating Markdown documentation with query examples into the root operations, the schema consumer gains a comprehensive understanding of how to effectively utilize the schema. This added clarity not only guides the consumer on the syntax for writing queries but also provides instructions on which specific queries to write and the overall business domain of the schema. As a result, the consumer can easily comprehend and navigate the schema, enhancing their ability to utilize it efficiently.

Summary

Documenting GraphQL code differs significantly from documenting code in other languages. It requires us to simultaneously consider both the individuals responsible for implementing the schema and the consumers of the schema. This dual focus ensures that our documentation effectively supports both parties, providing them with the necessary guidance and understanding to work with the GraphQL code efficiently.

In this chapter, we learned different documentation methods. We started by documenting types and interface fields, before moving on to input fields with a particular focus on highlighting the significance of documenting validation details.

Then, we ended up documenting queries and mutations, writing examples inside their documentation using the Markdown syntax to include GraphQL Query Language examples within the schema. We also documented piped queries and mutations with examples of their usage.

Now, we often find ourselves copying code between a type and its input, or between an interface and the type that implements it. At first, it may not be too difficult, but as the schema evolves, we need to remember to make changes in every part of the schema where that field is implemented. It's not an easy task, but fortunately, we have tools that allow us to visualize the schema and quickly observe these dependencies, which come to our aid. In the next chapter, we will explore them.

13
Tackling Schemas with Visualization

When working with GraphQL, understanding the schema is crucial. It serves as the backbone of your GraphQL API, defining the available types and fields, as well as their relationships. However, as your schema grows in size and complexity, it can become challenging to grasp its entirety and navigate through its intricacies. This is where schema visualization comes to the rescue.

Schema visualization provides a bird's eye view of your GraphQL schema, allowing you to quickly glance at an overview of your schema's structure and make informed decisions when designing and evolving your API. However, the benefits don't stop there. A visual representation of your schema also aids in communication and collaboration, providing a common language that developers, designers, and stakeholders can use to discuss and align on the API's capabilities. With a shared understanding of the schema, teams can work together more efficiently, reducing misunderstandings and accelerating development.

As GraphQL is a graph language, we can benefit from building schema files visually. Normally, GraphQL code is vertical, meaning most screen space of the coding IDE is left unused, like this:

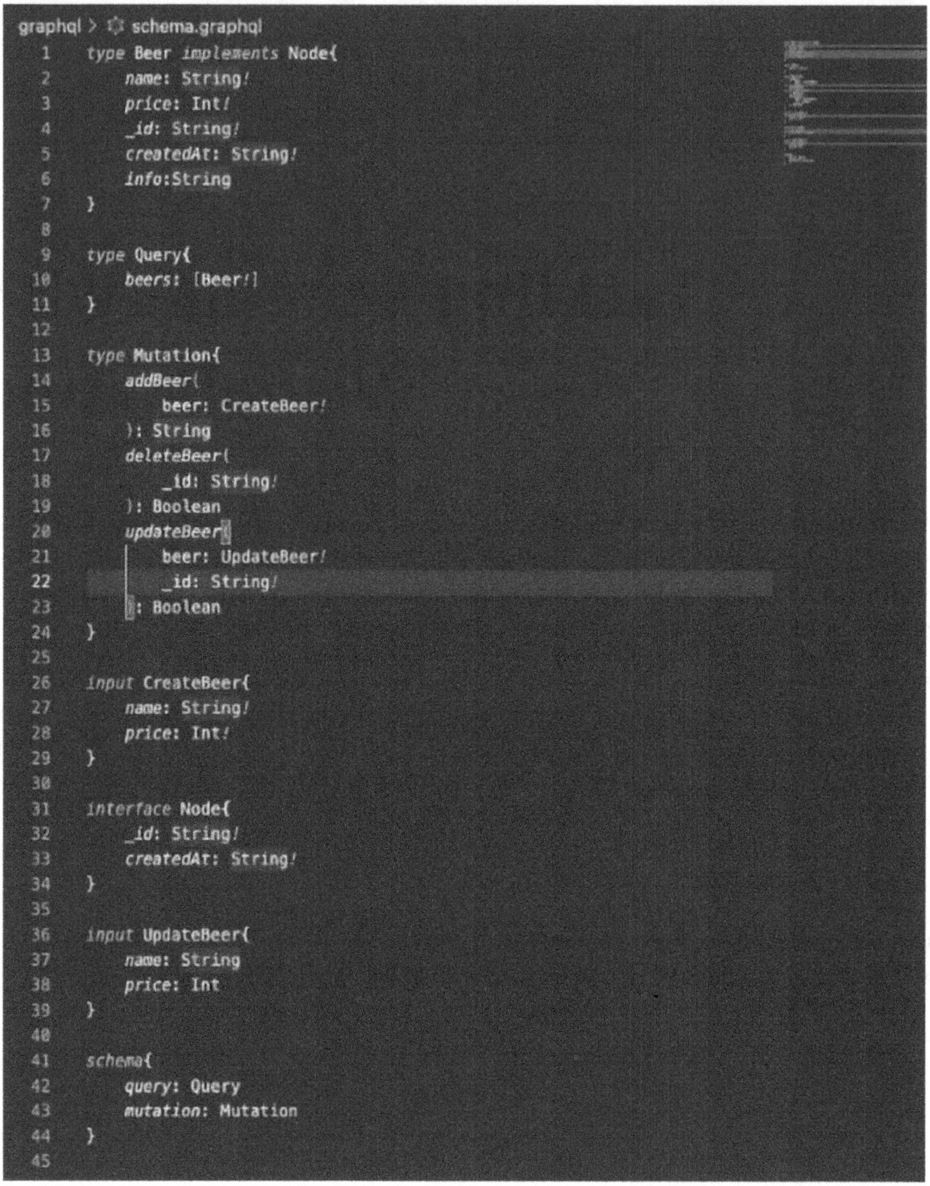

```
graphql >  schema.graphql
  1    type Beer implements Node{
  2        name: String!
  3        price: Int!
  4        _id: String!
  5        createdAt: String!
  6        info:String
  7    }
  8
  9    type Query{
 10        beers: [Beer!]
 11    }
 12
 13    type Mutation{
 14        addBeer(
 15            beer: CreateBeer!
 16        ): String
 17        deleteBeer(
 18            _id: String!
 19        ): Boolean
 20        updateBeer(
 21            beer: UpdateBeer!
 22            _id: String!
 23        ): Boolean
 24    }
 25
 26    input CreateBeer{
 27        name: String!
 28        price: Int!
 29    }
 30
 31    interface Node{
 32        _id: String!
 33        createdAt: String!
 34    }
 35
 36    input UpdateBeer{
 37        name: String
 38        price: Int
 39    }
 40
 41    schema{
 42        query: Query
 43        mutation: Mutation
 44    }
 45
```

Figure 13.1: Coding GraphQL using a programming IDE (don't worry about
reading all of the text; this is just an overview of the screen)

As you can see, the code occupies the left side of the screen space. However, through the utilization of visualizers, we can effectively utilize this space by representing an undirected graph that spans the entire screen. This way, we can see up to five times more GraphQL code at once, as well as the connections between the nodes.

> **Note**
>
> An **undirected graph** is a mathematical structure composed of a set of vertices or nodes, connected by edges that have no designated direction. In an undirected graph, the edges represent symmetric relationships between the vertices, allowing for bidirectional traversal and exploration.

In this chapter, you will learn what it means to onboard to a schema, as well as visualization techniques that will help you filter out the noise – these include filtering nodes and focusing on certain parts of the schema. You will also learn visual ways of working to find inconsistencies, repetitions, and security flaws inside your schemas.

By the end of this chapter, you will have a solid understanding of how to work with big schemas.

In this chapter, we will cover the following topics:

- Onboarding to the schema
- Improving existing schemas using the visual environment
- Using GraphiQL's navigation pane

Technical requirements

To complete this chapter, you will need the following:

- The open source GraphQL Editor (`https://github.com/graphql-editor/graphql-editor` and `https://graphqleditor.com`) will be needed. For this, run the following commands:

  ```
  git clone https://github.com/graphql-editor/graphql-editor
  npm i
  npm run run-all
  ```

  After running those commands, your browser should open with many examples visible. Choose the **Pure** example.

- You will need to be running one of the backends created inside the book to expose GraphiQL.

You can access the code in this chapter in the book's GitHub repository:

`https://github.com/PacktPublishing/GraphQL-Best-Practices/tree/main/chapter-13`

Onboarding to the schema

Onboarding refers to the process of familiarizing oneself with a large schema. It typically involves understanding the structure, relationships, and types within the schema. When onboarding to a big schema, one is often faced with a large number of lines of code, making it challenging to navigate and comprehend. The section aims to provide alternative ways to expedite the onboarding process by utilizing visualization techniques.

To start, you need to open your GraphQLEditor desktop app or use your GraphQL Editor cloud account, then clone the GraphQL Editor repository shown in the *Technical requirements* section.

Once done, copy the schema code from the chapter repository to the GraphQL Editor code pane. Then, on the right of your screen, you should see the following graph:

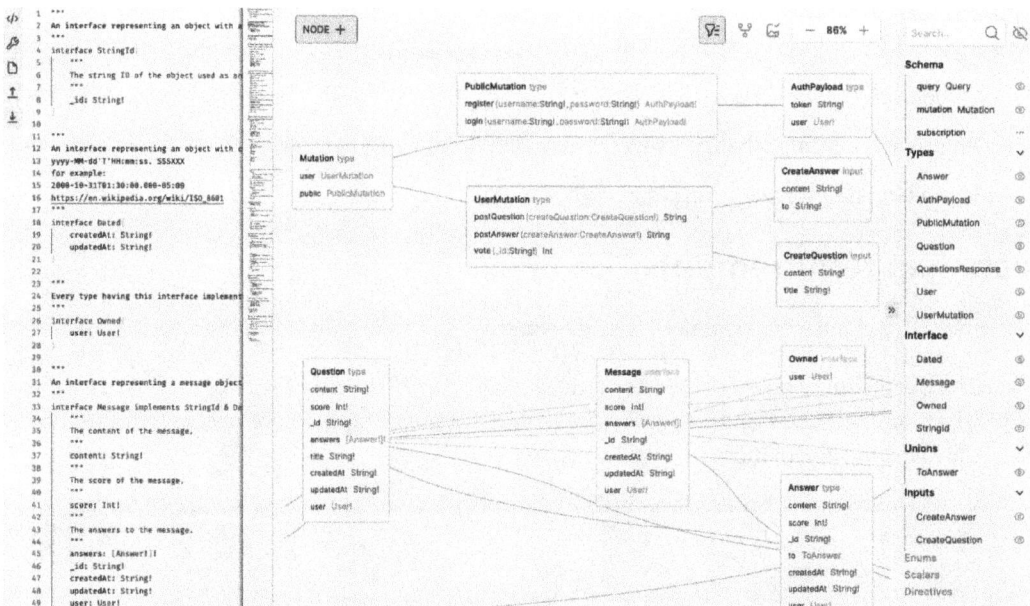

Figure 13.2: A full schema graph (don't worry about reading all
of the text; this is just an overview of the screen)

As you can see, there is a lot of information inside the schema, with the graph including many kinds of GraphQL nodes. We can limit the noise using a variety of visualization techniques, which we will cover now. The first involves hiding the node fields and only showing relations between the nodes.

Relations only

With the **Relations only** method, you can display the relationships between certain nodes. This way, you can understand how certain types connect through the graph.

To activate the **Relations only** mode, click the **Relations only** button at the top of GraphQL Editor:

Figure 13.3: Relations only button

Once clicked, all the fields inside the nodes should be hidden:

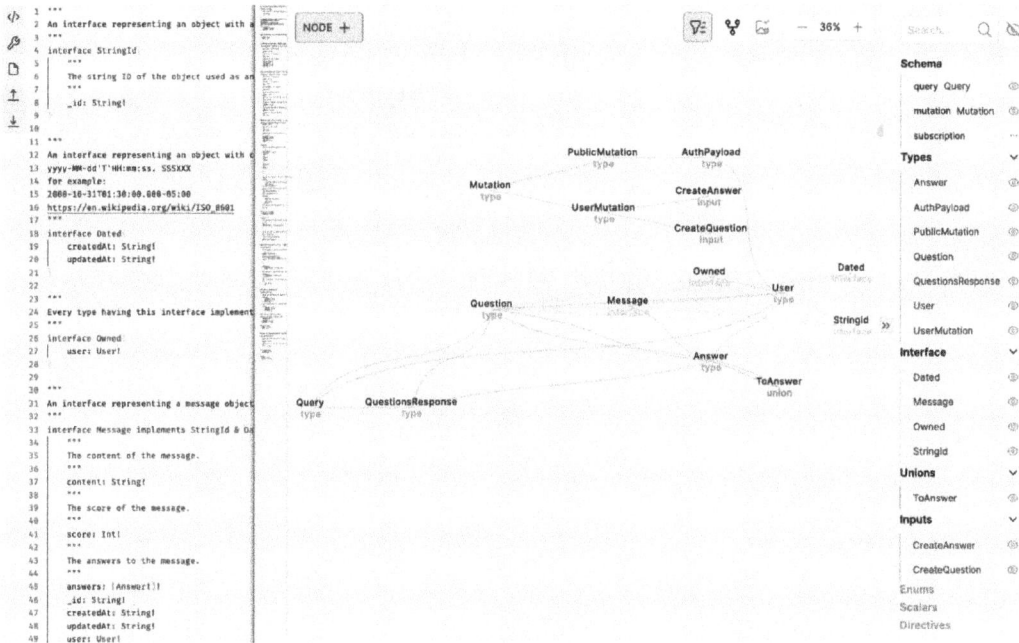

Figure 13.4: The graph in Relations only mode (don't worry about reading
all of the text; this is just an overview of the screen)

Now, we can't see the fields of our nodes and the titles of the nodes are bigger. Overall, we get a better picture of a schema, allowing us to see the following:

- Which nodes are related to each other

- Which inputs are used inside queries and mutations

- Which nodes implement interfaces

- What the union members are

So, by using **Relations only**, we can easily find the best path from one node to another.

Though we can see much more in the graph, we can also filter some nodes that we don't need now. Let's look at that next.

Filter by root type

To get a better picture of schema relations, we can disable some types. Depending on your situation, we won't need certain kinds of nodes:

- In most cases, it is unnecessary to represent enums and scalars as relations on the graph, as they are already visible and accessible as field types.

- We will only need input nodes inside the relation graph if our inputs depend on each other.

- If the schema is huge, we should definitely turn off everything except types and see the rest only in **Focus** mode (you will learn how to use **Focus** mode in the next section)

We will now turn off the input types – we only need those types when we are building the query, but for just seeing a bigger picture of a schema, they are not necessary.

To do this, click the **Filter by root type** button at the top menu of GraphQL Editor. Then, from the list, click the **input** button to disable input nodes rendering inside the graph (you can deselect any other nodes from here too):

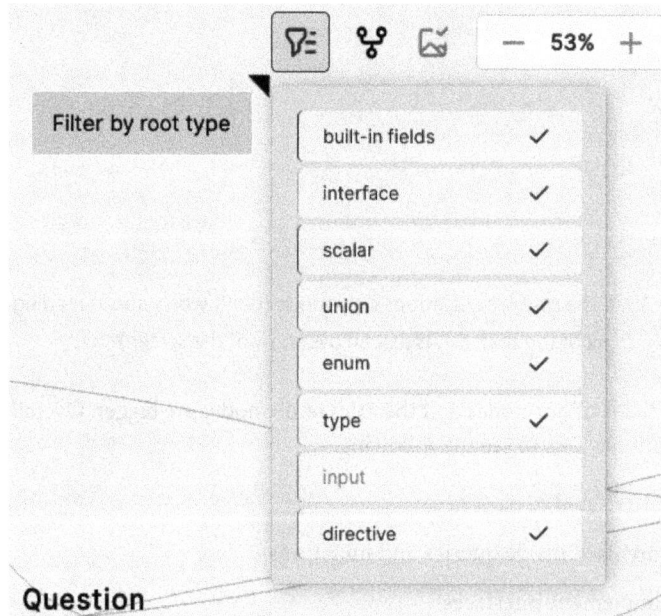

Figure 13.5: The filter menu

Now you will see the schema without input nodes:

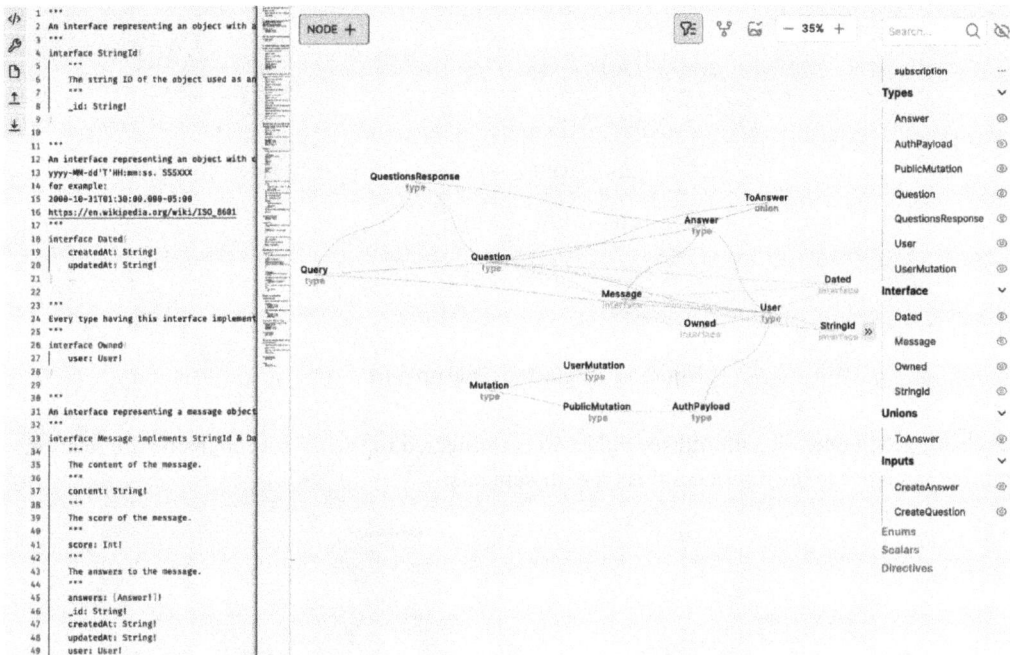

Figure 13.6: Schema without enums, scalars, and inputs (don't worry about
reading all of the text; this is just an overview of the screen)

Filtering certain types is helpful because we can see more inside one view. This way, we can have a
bigger picture of our schema that doesn't contain inputs used by the schema consumer.

Focusing on part of the schema

By utilizing graph visualization, you have the ability to easily navigate through relationships in order
to determine the path to a specific type before writing the query. By simply clicking on the field type,
you can effortlessly move to the associated type. This functionality allows you to efficiently determine
the connection between the root query type and any other desired type.

For example, when working with the User type, we can click on the **User** node, then click the **Focus**
button to focus on it and see all of its related types:

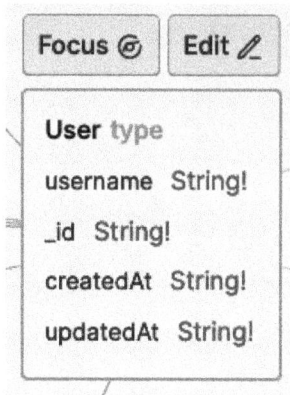

Figure 13.7: Selected node options

After clicking on **Focus**, we should see only nodes connected to the user node inside the relation graph. We can also see the interfaces this type implements:

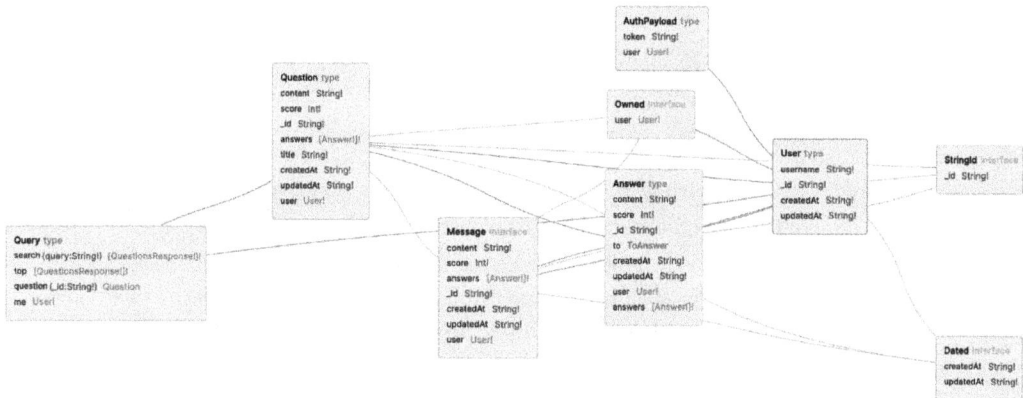

Figure 13.8: Focused User node inside the relation graph (don't worry about reading all of the text; this is just an overview of the screen)

Focusing on specific nodes may be useful when the schema consumer wants to select fields from different connected types during query composition. It is useful for both the schema consumer and the person who implements the backend as they can focus on one certain type at a time.

Traversing relations

Another useful visualization technique is **relation traversing**. This refers to the process of navigating through the relationships and connections between different types in a GraphQL schema. This allows developers to explore and understand the structure of the schema by visually traversing the relations between types.

To enter relation traversing, select a node and click the **Edit** button:

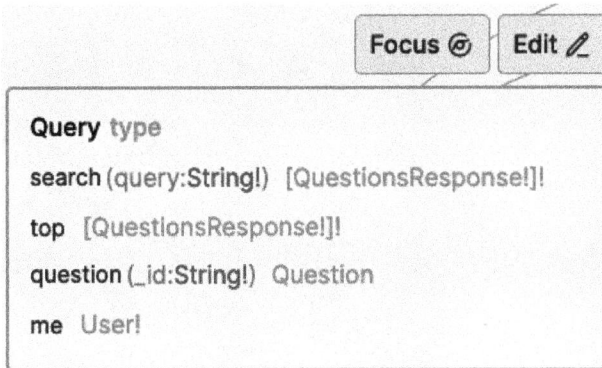

Focus ⊚   Edit ✎

Query type

search (query:**String!**) [QuestionsResponse!]!

top [QuestionsResponse!]!

question (_id:**String!**) Question

me User!

Figure 13.9: The selected Query node

You will now enter **Edit** mode, where you can further expand the types of fields:

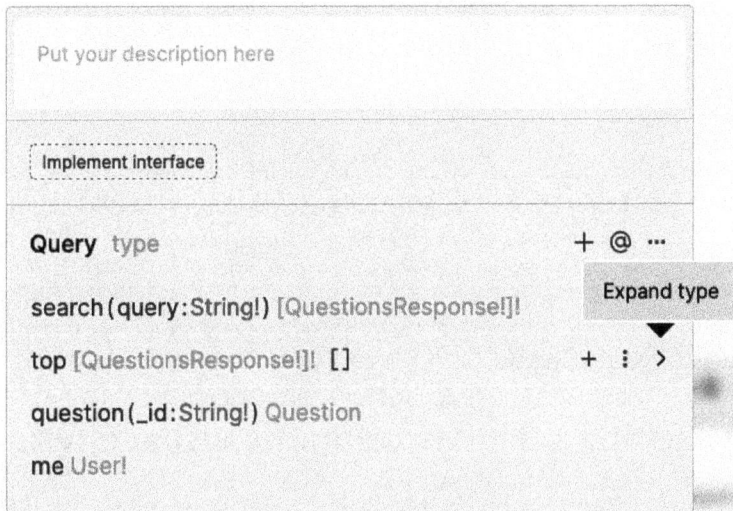

Put your description here

Implement interface

Query type   + @ ⋯

search (query:**String!**) [QuestionsResponse!]!

                                                    Expand type
top [QuestionsResponse!]! []   ▼

                                                    + ⋮ >
question (_id:**String!**) Question

me User!

Figure 13.10: The Query node's Edit mode

When you expand the fields, you should see the type of the field appearing as a separate stacked view on top of the current node:

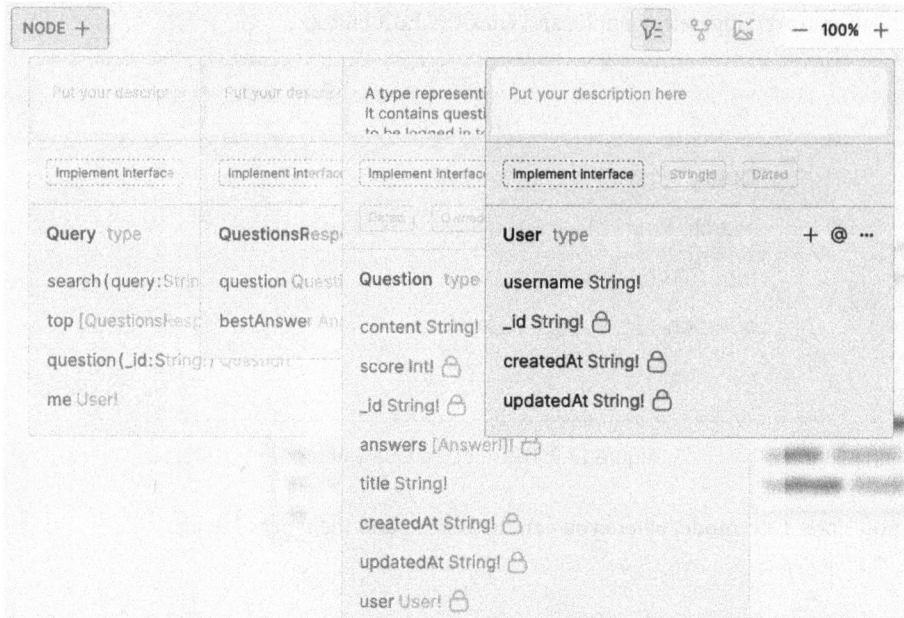

Figure 13.11: Nodes expanded in Edit mode

This functionality is particularly useful when writing queries, as it helps developers identify the correct path to retrieve the desired data. By visually traversing the relations, developers can efficiently determine the connection between the root query type and any other type, ensuring accurate and efficient data retrieval.

With that, we have analyzed several possible ways of using visualization to better understand large GraphQL schemas. In the next section, we will explore how we can leverage visual environments for building and editing GraphQL schemas.

Improving existing schemas using the visual environment

In my opinion, visualizers are a good tool to fix schemas – sometimes, what can't be seen directly inside the code can easily be seen through visualization. In this section, we will see how we can utilize visual ways of working to find inconsistencies, repetitions, and security flaws inside our schema.

Orphan nodes

Even on a big graph, we can see orphan nodes. These are nodes that we created and then forgot about. There are several reasons why this may happen:

- The sheer size and complexity of the schema can make it challenging to keep track of all the nodes. With numerous types, fields, and relationships, it becomes easier to overlook certain nodes that may not be immediately relevant or frequently used.

- Schemas evolve over time, with new nodes being added or existing nodes being modified. During these modifications, it is possible that certain nodes become obsolete or redundant. However, if these nodes are not properly reviewed or pruned from the schema, they can be forgotten and left unused.

- Human error and oversight can also contribute to forgetting about nodes in a large GraphQL schema. With multiple developers working on the schema, miscommunication or lack of documentation can lead to nodes being unintentionally overlooked or forgotten.

Luckily, the orphan nodes can easily be detected because they are separated and not connected to other nodes with any lines. Take a look at this schema example:

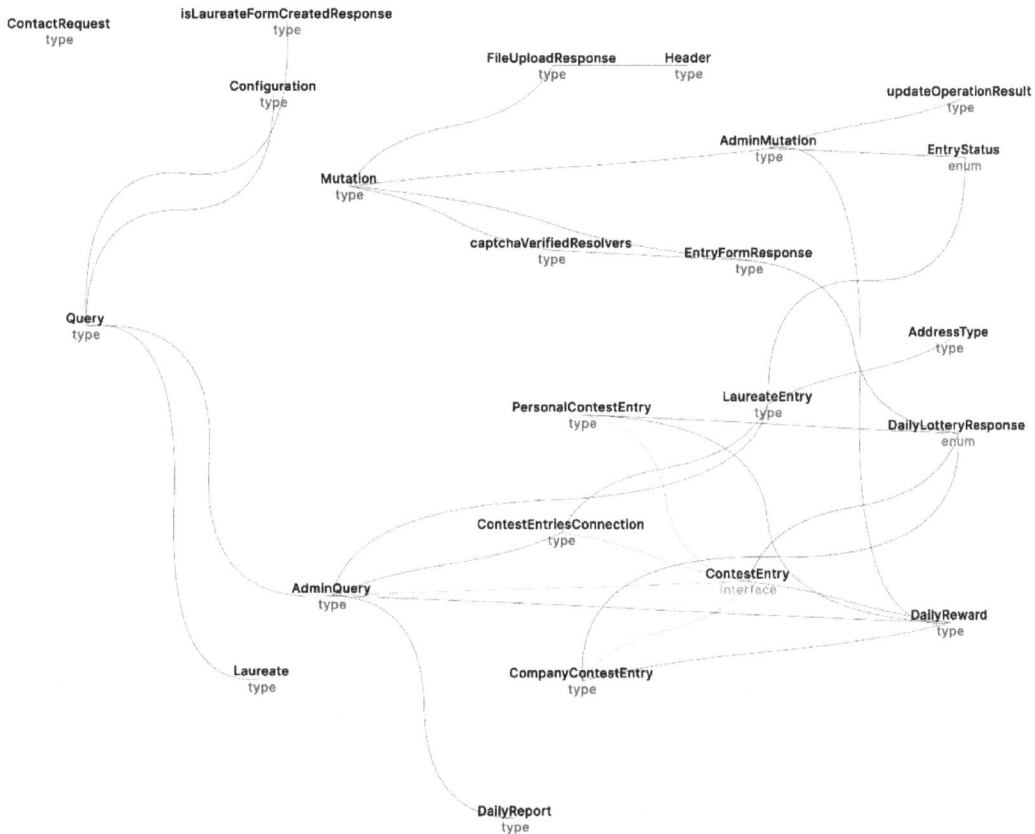

Figure 13.12: Example GraphQL schema visualized in GraphQL Editor (don't worry about reading all of the text; this is just an overview of the screen)

We can see that the **ContactRequest** node (in the top-left corner) is not used anywhere inside the schema and is not facing the schema consumer because it is not connected to the **Query**, **Subscription**, or **Mutation** nodes. That means we can delete it.

The availability of visualization techniques simplifies the process significantly, as it eliminates the need for manual searching and careful examination of each individual node's usage. For instance, if we encounter a node called **ContactRequest** and observe that it is utilized only once or not at all, we can confidently consider deleting it. Without visualization, however, the process would entail meticulously inspecting the schema node by node using a programming IDE.

Too many relations

Having multiple relationships within our schema can be beneficial as it allows us to access any object from various fields, regardless of their location. However, it is important to note that having an excessive number of relationships or objects that are interconnected with everything can have negative consequences.

When there are too many relationships, it can lead to a complex and convoluted schema, making it difficult to understand and maintain the schema. It may also result in performance issues, as fetching data becomes more time-consuming due to the extensive network of relationships.

By visually representing the schema and its relationships, we can easily spot relationship lines that intersect or cross with multiple nodes, indicating a high number of connections. This visual cue helps us quickly recognize areas of the schema that have a high degree of interconnectedness, like so:

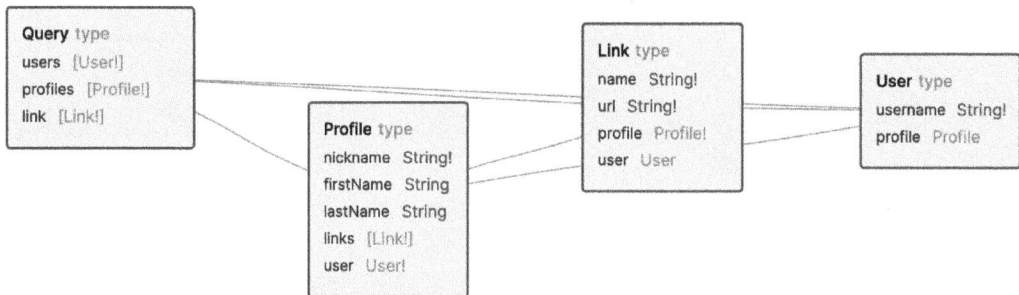

Figure 13.13: A schema with too many relations

Here, we can see a high degree of interconnectedness. However, some of these relationships are unnecessary. If we don't remove those unwanted connections, it can lead schema consumers to bad practices and over-fetching.

Having too many paths to the desired type can lead schema consumers to write queries like this one:

```
query GetUsers{
users{
   username
}
}
query GetProfiles{
```

```
profiles{
  nickname
  firstName
  lastName
  user{
    username
  }
}
}
query GetLinks{
  links{
    url
    name
    profile{
      user{
        username
      }
    }
  }
}
```

However, it is important to avoid relying on the frontend to match the returned data based on the schema, as this task should ideally be performed on the backend. By ensuring that the backend properly handles data matching based on the schema, we can maintain a more efficient and reliable data flow within the application.

> **Note**
>
> You should only be joining data from different sources if you can't modify the backend. These three queries could actually be one if the schema was better designed.

Looking back at *Figure 13.13*, if we already have a relation from the User node to the Profile node, and from the Link node to the Profile node, we can remove the relation between the Link and User nodes. This can be done by removing the user field from the Link type. Now look at *Figure 13.14* and how much cleaner the graph is:

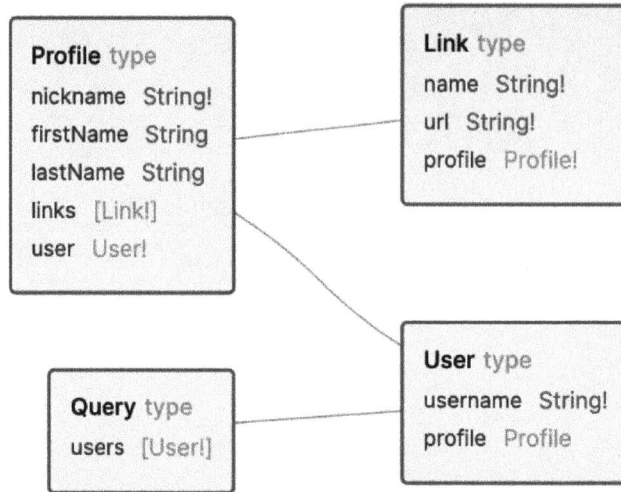

Figure 13.14: The schema after removing unwanted relations

This way, we can simplify the schema logic. Plus, the schema consumer will still have access to all the objects with the following query:

```
{
  Users{
    username
    profile{
      firstName
      lastName
      nickname
      links{
        name
        url
      }
    }
  }
}
```

This rule also applies to larger graphs – if you can see that the graph is clear and understandable without a lot of crossing lines, you can be sure that the relationships have been built efficiently.

Spotting security fails

Sometimes, we design schemas without looking at the visualization and we can make mistakes that can allow unauthorized access. Let's look at the following schema:

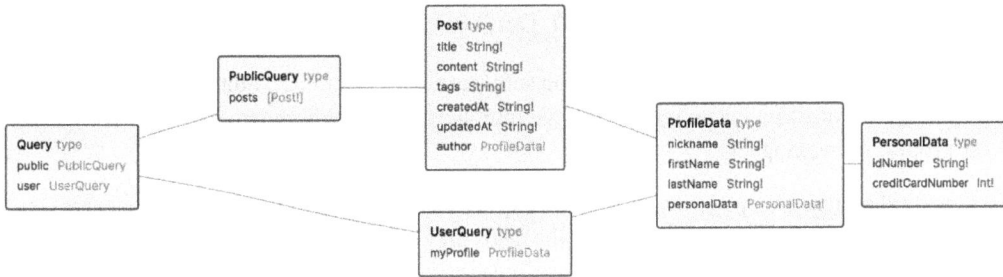

Figure 13.15: Schema with exposed PersonalData

Here, the `PersonalData` type is returned from `ProfileData.personalData`. However, it should be returned only for logged-in users, as it is behind the `UserQuery` authorization pipe and contains confidential data.

So why is it also returned from the `Post` type, which is returned from `PublicQuery`?! This can easily be spotted using visualization as two lines point to the `ProfileData` type. This way, all schema consumers that have access to public posts will have also access to every author's `PersonalData` type, which we don't want.

To fix this, we can create a new `AuthorProfile` type and the `Profile` interface that holds shared values of both the `AuthorProfile` and `ProfileData` types. This way, we will split the `ProfileData` type and avoid exposing the `PersonalData` type:

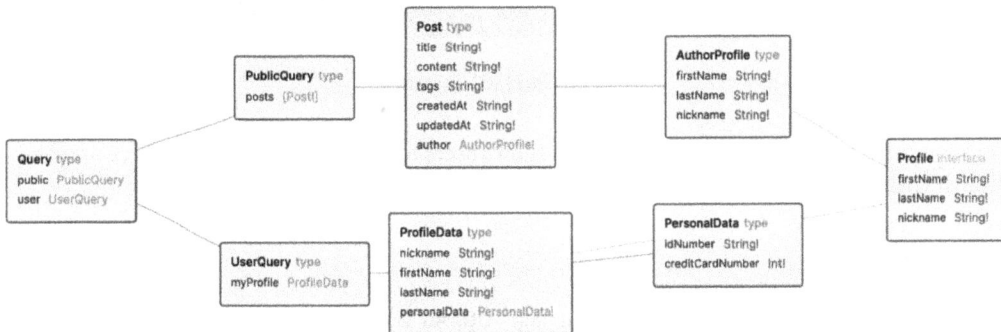

Figure 13.16: The schema after fixing the security issue

Now there is no connection between `PublicQuery` and `PersonalData`, making it impossible to get somebody's personal data from public queries.

So, as you can see, using these visualization techniques complements coding GraphQL schemas, helping us build consistent understandable schemas, detect unused nodes, and spot security fails.

In the final section, we will learn how to browse the schema using the GraphiQL interface.

Using GraphiQL's navigation pane

If we don't have time to dive into full visualization software such as GraphQL Editor, we can use some GraphiQL functions instead. Just to remind you, GraphiQL is usually built into most GraphQL servers in every programming language.

Using GraphiQL's navigation pane can save us in GraphQL when we are facing huge schemas. We can use the pane to just see the type names without the fields they implement. This is useful as we can find the desired type much faster than we would when traversing the whole schema code.

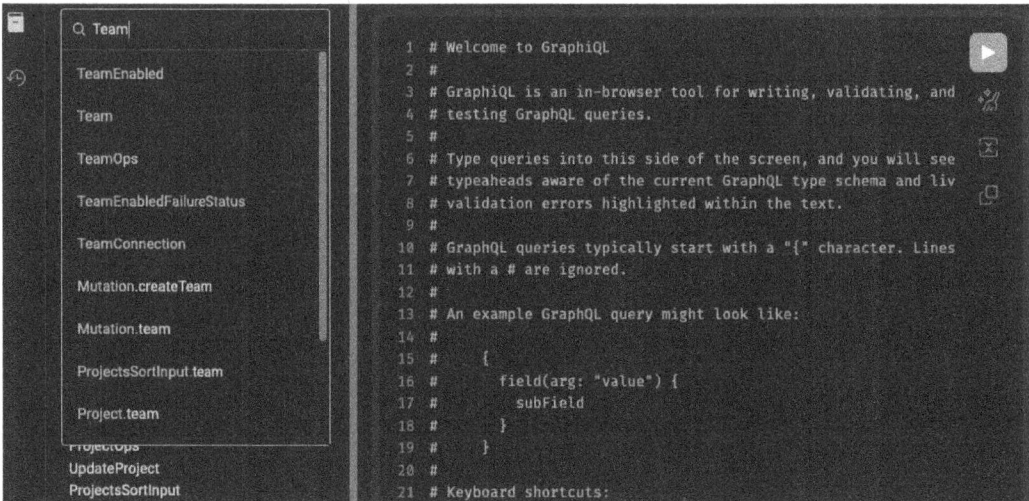

Figure 13.17: GraphiQL interface

It also has a really good fuzzy search engine that allows users to find relevant information even when the search terms are misspelled or imprecise.

Summary

To many people, working on the GraphQL language without visualization is almost impossible, and it is good to look at the visualization when working on a frontend, backend, or full-stack project. It just speeds up your work and makes problems easy to spot.

In this chapter, we learned the most useful visualization methods that help us to onboard to the schema. We saw how to filter nodes, focus only on a part of the schema that we need, and traverse graphs relation by relation. Then we moved on to understand how to modify and optimize our schemas, as well as fix security flaws. Without visualization, all of this would take a lot more time!

In the next chapter, we will use the knowledge gained thus far and put it into practice through a loyalty app. We will create the backend in the next chapter and the frontend in the final chapter.

Part 5 –
From an Idea to a
Working Project

In this part, we will create a booking system for a small business, such as a hairdressers. We will start with the GraphQL schema design, and then work our way up to create a backend and frontend for this system.

This part contains the following chapters:

- *Chapter 14, From an Idea to a Working Project – Backend Development with GraphQL and TypeScript*

- *Chapter 15, From an Idea to a Working Project – Frontend Integration with GraphQL and TypeScript*

14

From an Idea to a Working Project – Backend Development with GraphQL and TypeScript

In this chapter and the next, we will create a full stack application for small businesses, such as hairdressers, aesthetic medicine clinicians, beauticians, dentists, and so on. The goal of the application is to connect existing clients with the service owner and thus build their loyalty.

There are many applications on the market that allow small service-based businesses to advertise themselves, but in our application, we will add the ability for clients to earn points, discounts, and priority appointments for their loyalty.

As with every project in this book, we will start by designing the schema of our service. Then, we will write the backend code, and create integration tests to verify that our backend works.

In this chapter, we will cover the following topics:

- Creating a system concept and schema
- Implementing the application's backend
- Writing integration tests

So, let's get to work because small business owners are waiting!

Technical requirements

To complete this chapter, you will need a GraphQL IDE such as VS Code, Sublime Text, Notepad++, or GraphQL Editor.

You can access the code in this chapter in the book's GitHub repository: `https://github.com/PacktPublishing/GraphQL-Best-Practices/tree/main/chapter-14`.

Creating a system concept and schema

Before we start creating our GraphQL schema, it is good to know exactly what we want to create. To do this, we will use two methods – we will look at a visual diagram and we will create some user stories.

Here is the visual diagram:

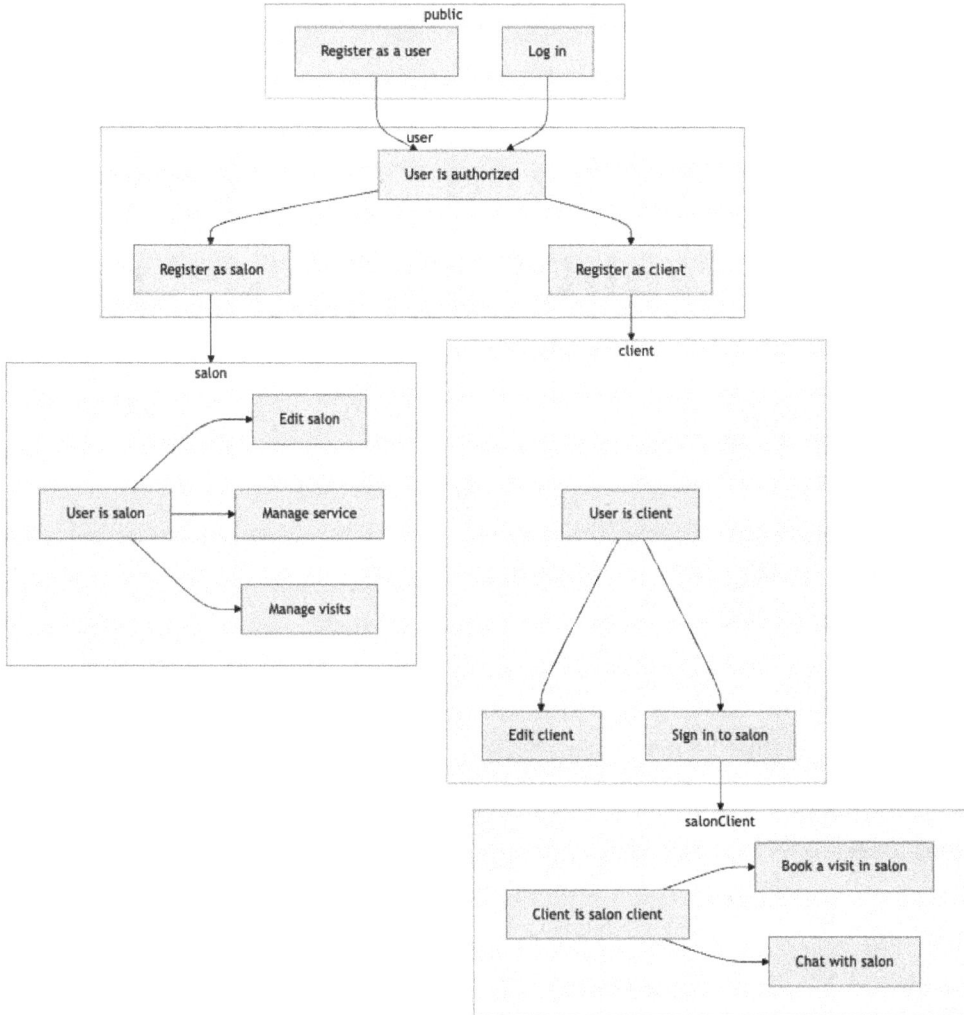

Figure 14.1: Visual diagram of the app

The diagram depicts key operations that the clients and salon owners will be able to complete, such as the ability to create and reserve services and chat with one another. We also need to remember that both the client and salon need to authenticate and authorize themselves.

Next, to make the creation of the schema clearer, we will write some user stories from the salon and client perspective. So, as a salon owner, we can write the following:

- I want to create a profile for my business on the application so that potential clients can discover and book appointments with me. This will help me keep my clients loyal.

- I want to manage my appointments through the application so that I can easily schedule, reschedule, or cancel appointments. This will help me stay organized and provide a better customer experience.

- I want to track my client's visits through the application so that I can reward them accordingly. This will incentivize repeat business and build customer loyalty.

- I want to have access to analytics and reports on my business performance through the application so that I can make data-driven decisions to improve my services and operations.

As a salon client, we can write the following:

- I want to create an account on the application so that I can easily book appointments with my preferred salon and receive benefits such as priority appointments and discounts

- I want to view my upcoming appointments so that I don't miss them and can plan my schedule accordingly

- I want to easily communicate with the salon owner or staff through the application so that I can ask questions, make special requests, or provide additional information before my appointment

Now we have a visual diagram of our app and some user stories, we can start crafting the GraphQL schema for our service.

Crafting the authentication and registration part of the schema

Every SaaS needs authentication and authorization so that we can identify salon owners and their clients. For this reason, we will start creating our GraphQL schema from the first user story – we need to allow the salon owner to create an account in our system.

To do this, we will use the same authentication and authorization mechanisms as in *Chapter 8*, so we can just copy the GraphQL responsible for that:

```
type AuthPayload{
  token: String!
  user: User!
}
type PublicMutation{
  register(
    username: String!
    password: String!
```

```
  ): AuthPayload!
  login(
    username: String!
    password: String!
  ): AuthPayload!
}
type Mutation{
  public: PublicMutation
}
```

Here, we have two methods – one to register the user account and one to log in. Inside our system, being a user means that you have the option to either register a salon or register as a client of a salon. `Mutation.public` is just a pipe for mutations that do not require authorization.

In order to fulfill the first story, we need to create mutations that will enable the creation of salons and clients:

```
input CreateSalon{
  name: String!
  slug: String!
}
input CreateClient{
  firstName: String!
  lastName: String!
  email: String
  phone: String
}
enum RegistrationError{
  EXISTS_WITH_SAME_NAME
  INVALID_SLUG
  INVALID_NAME
}
type UserOps{
  registerAsSalon(
    salon: CreateSalon!
  ): RegisterResponse
  registerAsClient(
    client: CreateClient!
  ): RegisterResponse
}
type RegisterResponse{
  errors: [RegistrationError!]!
}
type Mutation{
```

```
   public: PublicMutation
   user: UserOps
}
```

In the UserOps type, we allow the ability for a service or client to register to the system. If everything goes well during registration, we return nothing, but if something goes wrong, we return a response with errors described by the RegistrationError enum.

Additionally, we have included UserOps as a field in our root mutation to make it accessible in the server, as well as a method for users to register as clients within our system (although, at this stage, it does not yet specify registration as a client of a specific salon).

Base objects

As our schema needs to implement the booking system as well, we need to implement some base types. Let's break them down now.

Client

We will describe the Client type first, as this is the result of registration as a client:

```
type Client implements StringId & Dated{
   firstName: String!
   lastName: String!
   email: String
   phone: String
   user: User!
   _id: String!
   createdAt: String!
   updatedAt: String!
}
```

The Client profile contains basic client information, including the user field that points to the related User object.

We avoid adding anything about visits here for security and implementation purposes – after all, salons shouldn't see client visits in other salons.

SalonProfile

Next, here is the SalonProfile type GraphQL code:

```
type SalonProfile implements StringId & Owned & Dated{
   name: String!
   slug: String!
```

```
  _id: String!
  user: User!
  createdAt: String!
  updatedAt: String!
  services: [Service!]
}
```

This code implements the `StringId`, `Owned`, and `Dated` interfaces, as well as a `name` field and a `services` field to show the salon services. We didn't include the `clients` field here as it will be publicly visible, and we don't want one client to see other salon clients.

Visits should not be included as a field on `SalonProfile`, as this would require additional work from both schema architect developers and schema consumers. Instead, the only type that should be publicly visible as a connection is `Service`. Therefore, we can easily add the field of the `Service` type inside the `Visit` type.

Service

Now, let's look at the `Service` type, which represents one of the services offered by the salon owner:

```
type Service implements Dated & StringId{
  salon: SalonProfile!
  approximateDurationInMinutes: String!
  name: String!
  description: String!
  price: Int
  createdAt: String!
  updatedAt: String!
  _id: String!
}
```

The `Service` type starts with the `salon` field, which links back to `SalonProfile` – this is not useful right now, but it will be useful when resolving the `Visit` object, as the schema consumer needs the `Visit` object, the `Service` type connected to the visit, and the `SalonProfile` type that created the `Service` type.

Of course, our service also has `name`, `description`, and `price` fields, as well as an `approximateDurationInMinutes` field, which will be used to inform the client how much time they need.

SalonClient

Next, we need to connect the client with a salon. For that, we will create a `SalonClient` object:

```
type SalonClient implements StringId & Dated{
  salon: SalonProfile!
  visits(
    filterDates: DateFilter!
    salonId: String
  ): [Visit!]!
  _id: String!
  createdAt: String!
  updatedAt: String!
  client: Client!
  messageThread: MessageThread!
}
```

Besides the `client` and `salon` fields, we have the fields from the `StringId` and `Dated` interfaces.

Moreover, this type will be only available for authorized clients or salons, so we can safely include the `visit` and `messageThread` fields, too.

Message

Our app assumes there is just one message thread between a salon and a client. Here are the `MessageThread`, `MessageSender`, and `Message` types that describe this functionality within our GraphQL schema:

```
type Message implements Dated & StringId{
  createdAt: String!
  updatedAt: String!
  _id: String!
  sender: MessageSender!
  messageThread: MessageThread!
  message: String!
}

union MessageSender = SalonClient | SalonProfile

type MessageThread implements StringId & Dated{
  salonClient: SalonClient!
  messages: [Message!]!
  _id: String!
  createdAt: String!
  updatedAt: String!
}
```

Here is a brief explanation of the types:

- `MessageThread` holds the conversation between `SalonProfile` and `SalonClient`
- `MessageSender` is a union type that is used as a sender field inside `MessageThread`
- `Message` is the type responsible for a single message between the salon and the client inside `MessageThread`

Visit

The last type we must show here is `Visit`, which is responsible for holding booked visits:

```
enum VisitStatus{
  CREATED
  CONFIRMED
  CANCELED
  RESCHEDULED
  COMPLETED
}
type Visit implements StringId & Dated{
  _id: String!
  createdAt: String!
  updatedAt: String!
  service: Service!
  status: VisitStatus!
  whenDateTime: String!
  client: Client!
}
```

Our `Visit` object holds the information about the client's visit to the salon. It implements the `StringId` and `Dated` interfaces and is connected to a `Service` type via the `service` field, which means that every visit needs to have a service assigned.

The `Visit` object includes the following:

- The `status` field informs both the user and the salon about the current state of the visit. This state is defined within the `VisitStatus` enum.
- The `whenDateTime` field contains the ISO string representation of the visit's date and time.
- The `client` field represents the relationship to the `Client` type.

Salon operations

As we can now register a salon, we can start creating schemas for all the salon queries and mutations behind the salon authorization pipe. We will start with salon queries. Here is the `SalonQuery` type, which holds all the possible queries for an authorized salon owner:

```
type AnalyticsAmountPerDate{
  date: String!
  amount: Int!
}

type SalonAnalytics{
  visitsPerDay: [AnalyticsAmountPerDate!]!
  cashPerDay: [AnalyticsAmountPerDate!]!
}
input DateFilter{
  from: String!
  to: String
}
type SalonQuery{
  me: SalonProfile!
  clients: [SalonClient!]!
  visits(
    filterDates: DateFilter!
  ): [Visit!]!
  analytics(
    filterDates: DateFilter!
  ): SalonAnalytics
  client(
    _id: String!
  ): SalonClient
}
```

The `SalonQuery` type has access to the salon profile we created during the registration process. The first field, `me`, gives the salon owner access to the `SalonProfile` object, while the second field, `clients`, provides access to confidential data about the salon's clients.

Both the `visit` and `analytics` fields can be filtered by a `DateFilter` to reduce the amount of information requested through the app at a given time. Specifically regarding `analytics`, it will show the salon's analytics per day using `Int` numbers.

Now, we need to attach the `SalonQuery` query to the main query to be accessible by the schema consumer. For this to be accessible, the salon owner must be logged in with a `SalonProfile` object assigned:

```
type Query{
  user: UserQuery
}
type UserQuery{
  me: User!
  salon: SalonQuery
}
```

Now, `Query.user` will check whether the user is logged in, and `UserQuery.salon` will check whether the user has a salon created and whether it is connected to the user.

As we have crafted all the queries needed by the salon in GraphQL, we should implement some mutations to create and inject the data into our GraphQL server:

```
input CreateService{
  approximateDurationInMinutes: String!
  name: String!
  description: String!
  price: Int
}
input CreateVisitFromAdmin{
  whenDateTime: String!
  serviceId: String!
  userId: String!
}
input MessageInput{
  message: String!
}
type SalonOps{
  createService(
    service: CreateService!
  ): String
  serviceOps(
    _id: String!
  ): ServiceOps
  update(
    salon: UpdateSalon!
  ): RegisterResponse
  delete: Boolean
  createVisit(
    visit: CreateVisitFromAdmin!
```

```
    ): String
    visitOps(
      _id: String!
    ): VisitOps
    sendMessage(
      salonClientId: String!
      message: MessageInput!
    ): Boolean
  }

  type UserOps{
    registerAsSalon(
      salon: CreateSalon!
    ): RegisterResponse
    registerAsClient(
      client: CreateClient!
    ): RegisterResponse
    salon: SalonOps
  }
```

SalonOps.createService is responsible for the creation of the services that the salon provides.

Then, we can update the salon data with the update field and delete the salon with the delete one. The SalonProfile object is then passed as a source from another resolver whose return type is SalonOps.

We will use a similar pattern with the serviceOps field – the schema consumer needs to provide an _id field value of the service to modify or delete it.

The last field in SalonOps is the sendMessage field, which allows us to send messages to the client. To send a message, we need the SalonClient object's _id and the message content.

We also added the salon field to the UserOps type to expose SalonOps to logged-in users.

Then, we can write the actual ServiceOps type:

```
  input UpdateService{
    approximateDurationInMinutes: String
    name: String
    description: String
    price: Int
  }
  type ServiceOps{
    delete: Boolean
    update(
      service: UpdateService!
```

```
    ): Boolean
    }
```

The `ServiceOps` type also has `delete` and `update` fields; these fields have access to the `Service` type objects through the `serviceOps` field of the `SalonOps` type.

Finally, the same mechanism is used for visits – we have the field to create a visit and the `visitOps` field to modify or delete them:

```
input UpdateVisitFromAdmin{
  whenDateTime: String
  serviceId: String
  userId: String
}

type VisitOps{
  update(
    visit: UpdateVisitFromAdmin!
  ): VisitResponse
  delete: Boolean
}

enum VisitError{
  INVALID_DATE
}

type VisitResponse{
  errors: [VisitError!]!
}
```

Here, we named the inputs `UpdateVisitFromAdmin` and `CreateVisitFromAdmin` in particular because only the salon owner can modify and delete visits – that's why we included this information inside the name; it wouldn't make sense to give that kind of access to a salon client from a business perspective. The `VisitResponse` type is used to pass structured and informative errors to the schema consumer when there is an error on the visit update.

Now, we can take care of the client's side of the schema.

Client operations

As we know, our client needs to be able to register with different salons inside their interface. So, first, we need to create a mutation that allows registering to the salons:

```
input UpdateClient{
  firstName: String
```

```
    lastName: String
    email: String
    phone: String
}
type ClientOps{
  update(
    client: UpdateClient!
  ): RegisterResponse
  registerToSalon(
    salonSlug: String!
  ): Boolean
  salonClientOps(
    _id: String!
  ): SalonClientOps
}
```

With the `registerToSalon` field, we can connect the client to the salon.

Clients can also query their data and get the salons they registered to. Here is the code for `ClientQuery`:

```
type ClientQuery{
  clients: [SalonClient!]!
  me: Client!
  client(
    _id: String!
  ): SalonClient
}
```

With that, we can access the logged-in client data and all the connections with salons. Then, from `SalonClient`, we can get our visits and services booked in the salon.

We also need the `SalonClientOps` type to undertake actions in the name of salon clients:

```
input CreateVisitFromClient{
  whenDateTime: String!
  serviceId: String!
}
type SalonClientOps{
  createVisit(
    visit: CreateVisitFromClient!
  ): VisitResponse
  sendMessage(
    message: MessageInput!
  ): Boolean
}
```

For a salon client, they can create an appointment at the salon and communicate with the salon (we don't want the client to be able to modify the visit, as mentioned previously).

Now that we have created the schema, we can move on to actually building the application.

Implementing the application's backend

In this section, we will implement our app's backend using the Axolotl framework and GraphQL Yoga server technology. We will set up our project, and then write all the methods and resolvers using pipe mechanisms. Let's get started.

Setting up the project

To code our project, we will use the same backend stack we already used in previous chapters. We will start with bootstrapping the project:

1. Open your terminal and enter the following command:

    ```
    npx @aexol/axolotl create-yoga
    ```

2. Then, enter the project folder and install the required additional dependencies:

    ```
    npm i mongodb i-graphql jsonwebtoken
    npm i -D graphql-zeus @types/jsonwebtoken
    ```

    Here, mongodb will be used to connect to our database and store information, i-graphql will be used with graphql-zeus to ensure type safety, and jsonwebtoken will be used to secure access to our backend.

3. Then, we need to add some helper commands to the package.json file. Here is what the scripts part of our package.json file should look like:

    ```
    "scripts": {
      "start": "MONGO_URL=mongodb://localhost:27017/
        loyolo node lib/index.js",
      "build": "tspc",
      "watch": "tspc --watch",
      "typings": "axolotl build -m ./src/models.ts
        -s schema.graphql && zeus schema.graphql
        ./src -n",
      "inspect": "npm run build && MONGO_URL=
        mongodb://localhost:27017/loyolo axolotl inspect
        -s ./schema.graphql -r ./lib/resolvers.js",
    ```

```
  "dev": "nodemon || exit 1",
  "test": "jest",
  "mongo-dev": "docker run -p 27017:27017
    --name=loyolo mongo:latest"
}
```

Let's break this code down:

- The start command specifies the MONGO_URL environment variable and starts the server.

- Then, the build and watch commands transpile the project.

- The typings command generates the axolotl model definitions for type-safe resolver implementation. It also generates zeus typings, which will be used to create type-safe ORM for our code base.

- The nodemon command starts the hot-reload environment for our backend project. To fully work, nodemon needs to be configured in the nodemon.json file in the project's root. Here is the content for that file:

```
{
  "exec":"npm run typings && tspc || exit 1 &&
    npm run start",
  "ignore": [
    ".git",
    "node_modules/**/node_modules",
    "src/models.ts",
    "src/zeus"
  ],
  "watch":[
    "src",
    "schema.graphql",
    "package.json",
    ".env*"
  ],
  "ext":"ts,mts,cts,graphql,json,js"
}
```

We need to ensure that whenever a file with the extension specified by the ext key is modified inside the src folder, or any of the three other files mentioned inside the watch list (schema.graphql, package.json, and .env) is modified, the command specified in the exec key is executed.

Every time the file change is made, nodemon will kill the process and execute the exec command. This concept is known as hot reload. This allows us to develop the code without killing and restarting the server each time to see the live changes.

To prevent an infinite loop with `nodemon`, we have added a list of files to ignore. This is necessary because the `axolotl` and `zeus` commands generate files that would cause the files to change whenever the command typings are run.

- Back to the `package.json` file, we finally added the `mongo-dev` command to run the `mongodb` Docker container locally.

4. Now, copy the contents of the schema to the `schema.graphql` file inside the project root.

5. Run the following command in the **Terminal** tab, which will start the local MongoDB database on port `27017`:

```
npm run mongo-dev
```

6. Then, in a separate **Terminal** tab, run this command to start the `nodemon` instance:

```
npm run dev
```

Now we are done with the setup, we can begin developing the project.

Setting up ORM

Object-relational mapping (ORM) is a programming technique that allows developers to interact with a database using object-oriented paradigms. It simplifies the process of interacting with databases by abstracting away the database operations. Using ORM improves code readability and maintainability, and allows for easier database portability.

As we are using `i-graphql` for type-safe `mongodb` management, we need to set up the `src/orm.ts` file (if you don't have a `src` folder, please create one now). It will consist of all the model types and the index creation:

```
import { ModelTypes } from '@/src/zeus/index.js';
import { iGraphQL, MongoModel } from 'i-graphql';
import { ObjectId } from 'mongodb';

export type UserModel = MongoModel<ModelTypes['User']> & {
  passwordHash: string;
  salt: string;
};

export type SalonModel =
  Omit<MongoModel<ModelTypes['SalonProfile']>, 'services'>;
export type SalonClientModel =
  Omit<MongoModel<ModelTypes['SalonClient']>,
    'messageThread'>;
```

```
export type ClientModel = MongoModel<ModelTypes['Client']>;
export type VisitModel = MongoModel<ModelTypes['Visit']>;
export type ServiceModel =
  MongoModel<ModelTypes['Service']>;
export type MessageThreadModel =
  MongoModel<ModelTypes['MessageThread']>;
export type MessageModel =
  MongoModel<ModelTypes['Message']>;
```

Here, we began by using the MongoModel generic type to create types compatible with Mongo. This generic replaces the relation between types with string fields.

For example, we included the necessary properties, such as passwordHash and salt, in UserModel. This means that we don't expose those properties in GraphQL schema, but we have them in the database model.

In SalonProfile, we omitted the services field because the Service model already has a connection to the salon through its salon field, eliminating the need to store it in SalonModel. Similarly, in SalonClientModel, we excluded the messageThread field.

Now, we need to add our models to our ORM system. Here is how you can do it:

```
export const orm = async () => {
  return iGraphQL<
    {
      User: UserModel;
      Salon: SalonModel;
      SalonClient: SalonClientModel;
      Client: ClientModel;
      Visit: VisitModel;
      Service: ServiceModel;
      Message: MessageModel;
      MessageThread: MessageThreadModel;
    },
    {
      _id: () => string;
      createdAt: () => string;
      updatedAt: () => string;
    }
  >({
    _id: () => new ObjectId().toHexString(),
    createdAt: () => new Date().toISOString(),
    updatedAt: () => new Date().toISOString(),
```

```
    });
  };


  export const MongOrb = await orm();
```

We added all our models to our `i-graphql` engine, together with common fields such as `_id`, `createdAt`, `updatedAt`, and their generation methods.

Finally, we exported the `MongoOrb` variable, which holds our ORM system. We will import it into every resolver that needs access to the database.

At the end inside the `orm.ts` file, as mentioned, we will add some indexes to `mongodb` to make some fields unique:

```
  MongOrb('User').collection.createIndex(
    {
      username: 1,
    },
    { unique: true },
  );


  MongOrb('Client').collection.createIndex(
    {
      user: 1,
    },
    { unique: true },
  );


  MongOrb('Salon').collection.createIndex(
    {
      user: 1,
    },
    { unique: true },
  );
```

By using the `createIndex` calls, we instructed MongoDB to enforce uniqueness on the `username` field in the `User` collection. Additionally, we specified that the `user` field in both the `Client` and `Salon` collections should also be unique, ensuring that there are no duplicate values for these fields in the respective collections.

As we have our project bootstrapped together with schema, we can now start implementing the authentication and authorization part of the app.

Implementing authentication and authorization

To implement authentication and authorization, as previously mentioned, we will reuse the `src/auth.ts` file we created in *Chapter 8*. So, create a `src/PublicMutation.ts` file with the same contents as `chapter-08/qa-backend/src/PublicMutation.ts`.

We already described the user authentication process in *Chapter 8* too, but in this project, we will have some more authorization methods. Besides checking whether the user is logged in, we need to check whether the user has `Salon` or `Client` roles.

Before we start writing resolvers in pipe setup, we can create a helper file, `src/commonResolvers.ts`, to hold those authorization methods. Here is the start of the file's contents:

```
import { getUserOrThrow } from '@/src/auth.js';
import { MongOrb } from '@/src/orm.js';
import { GraphQLError } from 'graphql';
import { YogaInitialContext } from 'graphql-yoga';

export const commonUserResolver = async (yoga: [unknown,
unknown, YogaInitialContext]) => {
  const src = yoga[0] as { user: string };
  return MongOrb('User').collection.findOne({
    _id: src.user,
  });
};
```

We started with `commonUserResolver`, which is used to convert the user `_id` returned from the source to the `User` object. We will use this resolver to resolve the `User` object from the `user` fields included in `Client` and `SalonProfile`. This pattern can be used to retrieve any objects that have a primary key or other indexed fields within the source.

Then, the file moves on to the next resolver, `commonAuthUserResolver`:

```
export const commonAuthUserResolver = async (
  yoga: [unknown, unknown, YogaInitialContext]
) => {
  const authHeader = yoga[2].request.headers.get(
    'Authorization'
  );
  if (!authHeader)
    throw new GraphQLError(
      'You must be logged in to use this resolver'
    );
  return getUserOrThrow(authHeader);
};
```

This resolver extracts the Authorization header from the HTTP request and passes it to the getUserOrThrow method from src/auth.ts. This method will be used for our authorization pipes, Query.user and Mutation.user.

Finally, we create common resolvers for the client and salon, which detect whether the logged-in user has the related Client or Salon object:

```
export const commonClientResolver = async (yoga: [unknown,
unknown, YogaInitialContext]) => {
  const user = yoga[0] as UserModel;
  const client = await
  MongOrb('Client').collection.findOne({
    user: user._id,
  });
  if (!client) throw new GraphQLError('Forbidden!. Register
    as a Client');
  return client;
};

export const commonSalonResolver = async (
  yoga: [unknown, unknown, YogaInitialContext]
) => {
  const user = yoga[0] as UserModel;
  const salon = await MongOrb('Salon').collection.findOne({
    user: user._id,
  });
  if (!salon)
    throw new GraphQLError(
      'Forbidden!. Register as a Salon!'
    );
  return salon;
};
```

These resolvers verify whether the user is the owner of a client or salon account. If the user is not the owner, an error is thrown. If the user is the owner, the object is passed to the next resolver for further processing.

Next, we need to implement GraphQL pipes for the logged-in user. This includes the UserOps and UserQuery operations.

UserOps

All the resolvers of the UserOps fields assume that the user is logged in and they have an account. But in our system, it is not enough to register as a user; we need to provide them with a way to register either as a salon or client, or both. To do that, create src/UserOps.ts with the following content:

```
import { createResolvers } from '@/src/axolotl.js';
import {
  commonClientResolver,
  commonSalonResolver
} from '@/src/commonResolvers';
import { MongOrb, UserModel } from '@/src/orm.js';
import { RegistrationError } from '@/src/zeus/index.js';

export default createResolvers({
  UserOps: {
    // resolver code goes here
  },
});
```

Here is the basic skeleton for creating UserOps resolvers. We will include all the resolvers inside the UserOps value. Later on, we will add resolver fields inside the UserOps object.

For client and salon authorization, we will use the already implemented commonClientResolver and commonSalonResolver methods. Those methods will check whether the user has the proper account and pass this account object to the next resolver:

```
    client: commonClientResolver,
    salon: commonSalonResolver,
```

That's why we created those methods earlier – to reuse them across queries and mutations.

To make use of those methods, we will write a method to register as a salon, with a UserOps. registerAsSalon resolver:

```
    registerAsSalon: async (yoga, args) => {
      const s = MongOrb('Salon');
      const src = yoga[0] as UserModel;
      const SalonExists = await s.collection.findOne({
        $or: [
          {
            name: args.salon.name,
          },
          {
            slug: args.salon.slug,
```

```
      },
    ],
  });
  if (SalonExists) {
    return {
      errors:
        [RegistrationError.EXISTS_WITH_SAME_NAME],
    };
  }
  await s.createWithAutoFields(
    '_id',
    'createdAt',
    'updatedAt',
  )({
    ...args.salon,
    user: src._id,
  });
  return;
},
```

This resolver checks whether there is already a salon with the name provided by the user. If it exists, we will throw an error with `EXISTS_WITH_SAME_NAME`. If it doesn't, we create a `Salon` object related to the currently logged-in user.

Then, we can add a `registerClient` resolver:

```
registerAsClient: async (yoga, args) => {
  const s = MongOrb('Client');
  const src = yoga[0] as UserModel;

  if (args.client.email) {
    const EmailExists = await
    s.collection.findOne({
      $or: [
        {
          email: args.client.email,
        }             ],
    });
    if (EmailExists) {
      return {
        errors:
          [RegistrationError.EXISTS_WITH_SAME_NAME],
      };
    }
```

```
    }
    await s.createWithAutoFields(
      '_id',
      'createdAt',
      'updatedAt',
    )({
      ...args.client,
      user: src._id,
    });
    return;
  },
```

To register a client, we check whether there is a client with the same email . If it already exists, we return an error. If not, we create a `Client` instance.

UserQuery

Now, we need to implement authorization mechanisms on our `UserQuery` object. This is simpler than with `UserOps` as we only need to reuse our common resolvers. Here is the code for `src/UserQuery.ts`:

```
import { createResolvers } from '@/src/axolotl.js';
import {
  commonClientResolver,
  commonSalonResolver
} from '@/src/commonResolvers.js';
import { UserModel } from '@/src/orm';

export default createResolvers({
  UserQuery: {
    client: commonClientResolver,
    salon: commonSalonResolver,
    me: async (yoga) => {
      const src = yoga[0] as UserModel;
      return src;
    },
  },
});
```

For the `client` and `salon` fields, we just use common resolvers (`commonClientResolver` and `commonSalonResolver`) that return the associated models that we created before and the `UserQuery.me` resolver just returns the `User` object.

As we have the possibility to register, log in, and define ourselves as a salon and/or client, we can move on and implement dedicated methods for salon owners.

Methods for salon owners

A salon needs to create its services and define its prices. It also needs access to its client base and be able to see visits connected to its services and clients. Plus, as we know from user stories, the salon needs analytics to improve its service in the future.

We can describe these functionalities with GraphQL types. Here is the graph of types that `SalonQuery` has connected:

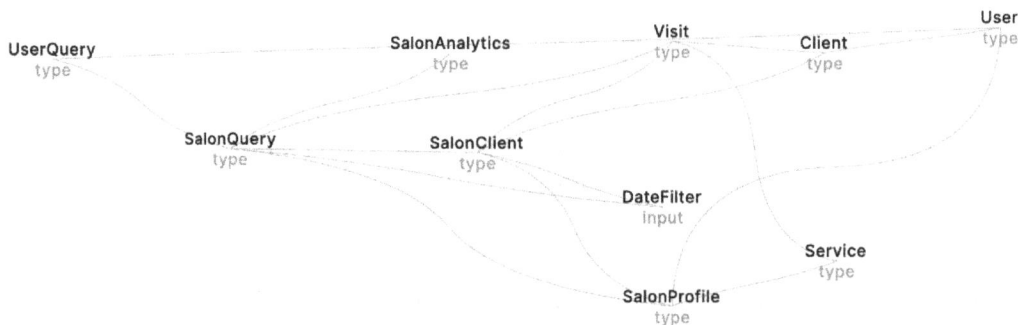

Figure 14.2: Graph including the SalonQuery path

As you can now see what the graph looks like, we can proceed to talk about all the GraphQL nodes. We will start by implementing queries and then we will implement mutations.

Queries

Our salon queries receive the `Salon` object from the resolver source and implement the read functionalities of a salon. Inside `SalonProfile`, we will implement methods to fetch the objects related to the salon.

SalonProfile

We will start coding salon queries from the `SalonProfile` fields. Here is the code responsible for `SalonProfile` located in `src/SalonProfile.ts`:

```
import { createResolvers } from '@/src/axolotl.js';
import { MongOrb, SalonModel } from '@/src/orm.js';

export default createResolvers({
  SalonProfile: {
    user: async (yoga) => {
      const src = yoga[0] as SalonModel;
      return MongOrb('User').collection.findOne({
        _id: src.user,
      });
```

```
      },
      services: async (yoga) => {
        const src = yoga[0] as SalonModel;
        return MongOrb('Service')
          .collection.find({
            salon: src._id,
          })
          .toArray();
      },
    },
  },
});
```

Inside this code, we return the User object for the SalonModel user field, which is a string holding an _id. In order to retrieve salon services, we can utilize SalonModel _id. By querying for all services with the salon field matching this _id, we can return all the matching services.

We also need to implement the Visit.service and Visit.client resolvers. Let's create a src/Visit.ts file with the following content:

```
import { createResolvers } from '@/src/axolotl.js';
import { MongOrb, VisitModel } from '@/src/orm.js';

export default createResolvers({
  Visit: {
    client: async (yoga) => {
      const src = yoga[0] as VisitModel;
      return MongOrb('Client').collection.findOne({
        _id: src.client,
      });
    },
    service: async (yoga) => {
      const src = yoga[0] as VisitModel;
      return MongOrb('Service').collection.findOne({
        _id: src.service,
      });
    },
  },
});
```

In both resolvers, we are looking for objects in the database. We are using VisitModel passed as a source of these resolvers. Then, we convert the IDs included in the client and service fields to the actual database objects.

Client

Next, create a `src/Client.ts` file with the following content:

```
import { createResolvers } from '@/src/axolotl.js';
import {
  commonUserResolver
} from '@/src/commonResolvers.js';

export default createResolvers({
  Client: {
    user: commonUserResolver,
  },
});
```

Here, just use the `commonUserResolver` function, which expects an object that has the `user` field of the `string` type, and then transforms it to the user object that is returned from the `Client.user` resolver.

Service

The `Service` model just links back to the salon profile. This will be beneficial for the client in the future, as they can easily access the salon information by following this path: `Visit` -> `Service` -> `SalonProfile`. But we can implement the resolver now. Here is how it should be written in `src/Service.ts`:

```
import { createResolvers } from '@/src/axolotl.js';
import { MongOrb, ServiceModel } from '@/src/orm.js';

export default createResolvers({
  Service: {
    salon: async (yoga) => {
      const src = yoga[0] as ServiceModel;
      return MongOrb('Salon').collection.findOne({
        _id: src.salon,
      });
    },
  },
});
```

Here, we used the `salon` field from the `ServiceModel` object to find the related salon and return it.

SalonClient

Now, we need to implement the `SalonClient` type – it is our glue between the authorized salon and authorized client, providing access to the client's visits to a particular salon. Here is the code for the resolvers located inside the `src/SalonClient.ts` file:

```
import { createResolvers } from '@/src/axolotl.js';
import { MongOrb, SalonClientModel } from '@/src/orm.js';

export default createResolvers({
  SalonClient: {
    client: async (yoga) => {
      const src = yoga[0] as SalonClientModel;
      return MongOrb('Client').collection.findOne({
        _id: src.client,
      });
    },
    salon: async (yoga) => {
      const src = yoga[0] as SalonClientModel;
      return MongOrb('Salon').collection.findOne({
        _id: src.salon,
      });
    },
    visits: async (yoga) => {
      const src = yoga[0] as SalonClientModel;
      return MongOrb('Visit')
        .collection.find({
          client: src.client,
        })
        .toArray();
    },
    messageThread: async (yoga) => {
      const src = yoga[0] as SalonClientModel;
      return MongOrb('MessageThread').collection.findOne({
        salonClient: src._id,
      });
    },
  },
});
```

Every user that gets access to the `SalonClient` resolvers is for sure logged in, and either has a salon themselves or is a client of a salon. In every resolver, we then just return objects based on the `src` variable with a source of the `SalonClientModel` type.

This process is necessary because, in our database, fields representing types are stored as `String` type IDs to establish relationships. To work with these relationships, we convert the `String` IDs into their corresponding object representations.

SalonAnalytics

The `SalonAnalytics` type is responsible for returning analytics data from every field it has, as well as gathering information from the parent resolver about the date and the `Salon` object. Here is the code for the `src/SalonAnalytics.ts` file:

```
import SalonQuery from '@/src/SalonQuery.js';
import { createResolvers } from '@/src/axolotl.js';
import { MongOrb } from '@/src/orm.js';
import type { SourceInfer } from '@/src/sourceInfer.js';

const aggregateByField = <T>(
  objects: T[],
  aggregationKeyFn: (o: T) => string,
  aggregationValueFn?: (o: T) => number,
) => {
  const aggregatedDict: Record<string, number> = {};
  objects.forEach((o) => {
    const keyName = aggregationKeyFn(o);
    if (!aggregatedDict[keyName]) {
      aggregatedDict[keyName] = 0;
    }
    aggregatedDict[keyName] += aggregationValueFn ?
      aggregationValueFn(o) : 1;
  });
  return Object.entries(aggregatedDict).map(([dateKey,
    amount]) => ({ date: dateKey, amount }));
};
```

Before we write the resolvers' code, here, we use the `aggregateByField` function (I am a TypeScript generic types fan so using this function will help you create all the analytics needed with less code). The function takes an array of objects of the `T` generic type that is inferred from the `objects` array argument.

> **Note**
>
> TypeScript generic types are a powerful feature that allows us to create reusable and flexible code by defining placeholders for types that can be specified later. They enable us to write functions and classes that can work with a variety of types, making our code more generic and adaptable.

Then, we specify `aggregationKeyFn`, which takes the object from the list of objects and extracts the key value from it so it can be used as a group index. This is needed to group the objects, for example, by date. In the end, we sum up values using `aggregateValueFn`, which extracts the value from the object in a list. If the function is not set, we just count the objects.

With this function in place, we can now actually implement the resolvers. Let's look at the end of the `src/SalonAnalytics.ts` file:

```
export default createResolvers({
  SalonAnalytics: {
    cashPerDay: async (yoga) => {
      const src = yoga[0] as SourceInfer<typeof SalonQuery>
        ['SalonQuery']['analytics'];
      const { from, to } = src.args.filterDates;
      const visits = await MongOrb('Visit')
        .collection.find({
          whenDateTime: {
            $gte: from,
            ...(to ? { $lte: to } : {}),
          },
        })
        .toArray();
      const relatedServices = await
        MongOrb('Visit').related(visits, 'service',
          'Service', '_id');
      return aggregateByField(
        visits,
        (o) => o.whenDateTime.slice(0, 10),
        (o) => relatedServices.find((rs) => rs._id ===
          o.service)?.price || 0,
      );
    },
    visitsPerDay: async (yoga) => {
      const src = yoga[0] as SourceInfer<typeof SalonQuery>
        ['SalonQuery']['analytics'];
      const { from, to } = src.args.filterDates;
      const visits = await MongOrb('Visit')
        .collection.find({
          whenDateTime: {
            $gte: from,
            ...(to ? { $lte: to } : {}),
          },
        })
        .toArray();
```

```
      return aggregateByField(visits, (o) =>
        o.whenDateTime.slice(0, 10));
    },
  },
});
```

Inside the `cashPerDay` resolver, we take all the visits within the range provided by the schema consumer. Then, using our `aggregateByField` function, we group them by date by slicing the first 10 characters from the `whenDateTime` ISO string. After that, we provide an unnamed arrow function to take the price value from the related object.

Meanwhile, inside the `visitsPerDay` resolver, we didn't have to provide the `aggregationValueFn` function as we just counted the objects.

SalonQuery

`SalonQuery` is our main type, containing connections to all other queries and also providing a resolver for salon visits. Here is the code of `src/SalonQuery.ts`:

```
import { createResolvers } from '@/src/axolotl.js';
import { MongOrb, SalonModel } from '@/src/orm.js';

export default createResolvers({
  SalonQuery: {
    analytics: async (yoga, args) => {
      const src = yoga[0] as SalonModel;
      return {
        model: src,
        args,
      };
    },
    clients: async (yoga) => {
      const src = yoga[0] as SalonModel;
      return MongOrb('SalonClient')
        .collection.find({
          salon: src._id,
        })
        .toArray();
    },
    client: async (yoga, args) => {
      const src = yoga[0] as SalonModel;
      return MongOrb('SalonClient').collection.findOne({
        _id: args._id,
        salon: src._id,
```

```
      });
    },
    visits: async (yoga, args) => {
      const src = yoga[0] as SalonModel;
      const services = await MongOrb('Service')
        .collection.find({
          salon: src._id,
        })
        .toArray();
      return MongOrb('Visit')
        .collection.find({
          service: {
            $in: services.map((s) => s._id),
          },
          whenDateTime: {
            $gte: args.filterDates.from,
            ...(args.filterDates.to
              ? {
                  $lte: args.filterDates.to,
                }
              : {}),
          },
        })
        .toArray();
    },
    me: async (yoga) => {
      const src = yoga[0] as SalonModel;
      return src;
    },
  },
});
```

Let's break down the resolvers:

- In the me resolver, we simply return the `src` object to enable the user to enter their SalonProfile type.

- In the clients field resolver, we return all the SalonClient objects that are associated with the respective salon.

- In the client field resolver, we can also get the SalonClient object using its _id. It is important to note that when performing this action, we need to filter the results based on the salon _id to prevent one salon from gaining access to another salon's client by their ID.

- The `visits` field resolver is more complicated. As we already have a connection between a service and a visit, and between a salon and a service, but we don't have a connection between a salon and a visit, we need to aggregate the data and properly filter it to find all the salon visits. Of course, we can access the `visit` object from the `SalonClient` resolvers, but to make things easier for the schema consumer, we will add this additional resolver. Inside this resolver, we also force the schema consumer to filter visits by date.

- Finally, the `analytics` field resolver takes the source and arguments and passes them down to the `SalonAnalytics` field resolvers. It is done like this to avoid having the same argument on every field of the `SalonAnalytics` type.

After we implement the methods to query the data, we need to implement the mutations.

Mutations

Salon mutations will contain all the operations that the salon can perform in their name. In the following graph, we can see all the GraphQL types associated with salon mutations:

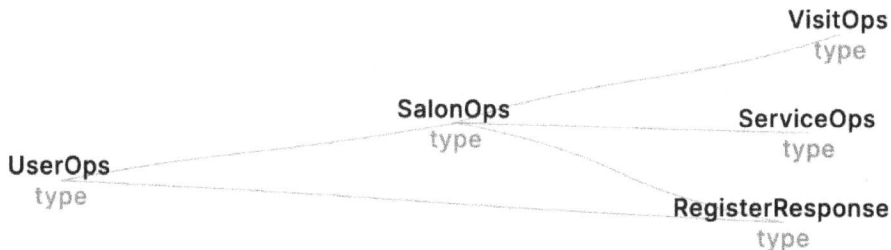

Figure 14.3: Salon mutations

As you can see, the graph depicts the operations of the `Service` and `Visit` objects. We will code resolvers to manage these salon services and visits.

VisitOps

Our `VisitOps` resolver is a simple resolver used to update or delete visits. Here is the content of the `src/VisitOps.ts` file:

```
import { createResolvers } from '@/src/axolotl.js';
import { MongOrb, VisitModel } from '@/src/orm.js';

export default createResolvers({
  VisitOps: {
    delete: async (yoga) => {
      const src = yoga[0] as VisitModel;
      const result = await
        MongOrb('Visit').collection.deleteOne(
```

```
        {
          _id: src._id,
        });
      return !!result.deletedCount;
    },
    update: async (yoga, args) => {
      const src = yoga[0] as VisitModel;
      await MongOrb('Visit').collection.updateOne(
        {
          _id: src._id,
        },
        {
          $set: {
            ...args.visit,
          },
        },
      );
    },
  },
});
```

In both resolvers, we take the visit `_id` from the source and pass it to either the `updateOne` or `deleteOne` function of mongodb.

ServiceOps

Now, let's take a look at the `ServiceOps` resolver code, located inside `src/ServiceOps.ts`:

```
import { createResolvers } from '@/src/axolotl.js';
import { MongOrb, ServiceModel } from '@/src/orm.js';

export default createResolvers({
  ServiceOps: {
    delete: async (yoga) => {
      const src = yoga[0] as ServiceModel;
      const result = await
        MongOrb('Service').collection.deleteOne(
        {
          _id: src._id,
        });
      return !!result.acknowledged;
    },
    update: async (yoga, args) => {
      const src = yoga[0] as ServiceModel;
```

```
      const result = await
        MongOrb('Service').collection.updateOne(
          {
            _id: src._id,
          },
          {
            $set: {
              ...args.service,
            },
          },
        );
      return !!result.acknowledged;
    },
  },
});
```

ServiceOps contains the delete and update resolvers that use the source passed to them, and then using the provided arguments and MongoDB functions, they perform the associated operation. Both resolvers return boolean if the object was updated/deleted.

SalonOps

SalonOps is the main set of operations that can be performed by an authorized salon. The structure of the src/SalonOps.ts file looks like this:

```
import { createResolvers } from '@/src/axolotl.js';
import { MongOrb, SalonModel } from '@/src/orm.js';
import { VisitStatus } from '@/src/zeus/index.js';
import { GraphQLError } from 'graphql';

export default createResolvers({
  SalonOps: {
    // resolver code goes here
  },
});
```

The skeleton is the same as in other files, but now we will go through the resolvers implemented inside SalonOps one by one. The first resolver is createVisit:

```
createVisit: async (yoga, args) => {
    const src = yoga[0] as SalonModel;
    const Service = await
      MongOrb('Service').collection.findOne(
        {
          salon: src._id,
```

```
          _id: args.visit.serviceId,
      });
      if (!Service) throw new GraphQLError(
        'Forbidden! Cannot create visit for other salon');
      const result = await
        MongOrb('Visit').createWithAutoFields(
          '_id',
          'createdAt',
          'updatedAt',
        )({
          service: args.visit.serviceId,
          client: args.visit.clientId,
          whenDateTime: args.visit.whenDateTime,
          status: VisitStatus.CREATED,
        });
      return result.insertedId;
    },
```

We start by checking whether the service provided by the schema consumer exists. If it does not exist, we throw an error; if it does exist, we create a visit.

The next resolvers are update and delete:

```
  delete: async (yoga) => {
    const src = yoga[0] as SalonModel;
    const result = await
      MongOrb('Salon').collection.deleteOne(
        {
          _id: src._id,
        });
    return !!result.acknowledged;
  },
  update: async (yoga, args) => {
    const src = yoga[0] as SalonModel;
    const result = await
      MongOrb('Salon').collection.updateOne(
        {
          _id: src._id,
        },
        {
          $set: {
            ...args.salon,
          },
        },
```

```
    );
    return !!result.acknowledged;
  },
```

We don't need to discuss them as they use the same pattern as other objects inside this backend (they are passed the source, and then use the provided arguments and MongoDB function), but they are used to delete and update our salon object.

Next, we will write resolvers related to services and visits associated with `SalonProfile`:

```
createService: async (yoga, args) => {
  const src = yoga[0] as SalonModel;
  const result = await
    MongOrb('Service').createWithAutoFields(
      '_id',
      'createdAt',
      'updatedAt',
    )({
      ...args.service,
      salon: src._id,
    });
  return result.insertedId;
},
serviceOps: async (yoga, args) => {
  const src = yoga[0] as SalonModel;
  return MongOrb('Service').collection.findOne({
    _id: args._id,
    salon: src._id,
  });
},
visitOps: async (yoga, args) => {
  const src = yoga[0] as SalonModel;
  return MongOrb('Visit').collection.findOne({
    _id: args._id,
    salon: src._id,
  });
},
```

As a reminder, inside the `createService` resolver, we are creating a `Salon` service, and inside the `visitOps` resolver, we are looking for the related `Visit` object inside the database to pass it to the next resolver.

The last resolver inside `SalonOps` is `sendMessage`, used to send messages to salon clients:

```
sendMessage: async (yoga, args) => {
  const src = yoga[0] as SalonModel;
  const thread = await
  MongOrb('MessageThread').collection.findOneAndUpdate(
    {
      salon: src._id,
      client: args.salonClientId,
    },
    {
      $setOnInsert: {
        salon: src._id,
        client: args.salonClientId,
      },
    },
    {
      upsert: true,
    },
  );
  if (!thread) throw new GraphQLError('Corrupted
    message thread. Please try again');
  const result = await
    MongOrb('Message').createWithAutoFields(
      '_id',
      'createdAt',
      'updatedAt',
    )({
      messageThread: thread._id,
      sender: src._id,
      message: args.message.message,
    });
  return !!result.insertedId;
},
```

First, we check whether there is an existing `MessageThread` object for the salon and client. If no such object exists, we proceed to insert it into the database. We do this using the `$setOnInsert` MongoDB feature and `upsert`. In the event that we are unable to successfully insert the object, an error is thrown. After ensuring the presence of the `MessageThread` object, we proceed to create the message itself.

At this point, we have implemented all the salon queries and mutations and can proceed to the salon client operations.

Methods for clients

Here, we will implement all the methods accessible from the client's perspective. We will start by writing queries to list the related salons and fetch information about themselves. Then, we will implement mutations to update the details of clients and for them to join salons. Let's get started.

Queries

In our client queries, our main focus is to fetch salons that clients have subscribed to. Since we have already written the `SalonClient` resolvers, there is only a small amount of code related to the client in this context.

Regarding `ClientQuery`, it is only used to get the client details and return all the connected salons. Here is the content of the `src/ClientQuery.ts` file:

```
import { createResolvers } from '@/src/axolotl.js';
import { ClientModel, MongOrb } from '@/src/orm.js';

export default createResolvers({
  ClientQuery: {
    me: async (yoga) => {
      const src = yoga[0] as ClientModel;
      return src;
    },
    clients: async (yoga) => {
      const src = yoga[0] as ClientModel;
      return MongOrb('SalonClient')
        .collection.find({
          client: src._id,
        })
        .toArray();
    },
    client: async (yoga, args) => {
      const src = yoga[0] as ClientModel;
      return MongOrb('SalonClient').collection.findOne({
        _id: args._id,
        client: src._id,
      });
    },
  },
});
```

The query contains two resolvers. In the me resolver, we simply pass the object that we received from `Query.client`. The `clients` resolver, on the other hand, is responsible for returning all the salons that the client registered to.

Additionally, we have the capability to query a specific salon client by using its `_id`. However, it is crucial to remember to apply a filter based on the client as well, in order to avoid retrieving someone else's `SalonClient` object.

Mutations

Now, we will start writing client and salon client mutations.

ClientOps

`ClientOps` is the main set of operations that can be performed by an authorized client. The structure of the `src/ClientOps.ts` file looks like this:

```
import { createResolvers } from '@/src/axolotl.js';
import { ClientModel, MongOrb } from '@/src/orm.js';
import { RegistrationError } from '@/src/zeus/index.js';
import { GraphQLError } from 'graphql';

export default createResolvers({
  ClientOps: {
    // resolver code goes here
  },
});
```

This is the basic skeleton for ClientOps resolvers.

Now, we will implement those resolvers inside it. First, we implement the `update` resolver:

```
update: async (yoga, args) => {
    const src = yoga[0] as ClientModel;
    if (args.client.email || args.client.phone) {
      const exists = await
      MongOrb('Client').collection.findOne({
        $or: [
          {
            phone: args.client.phone,
          },
          {
            email: args.client.email,
          },
        ],
      });
      if (exists) {
        if (exists._id !== src._id)
        return [RegistrationError.EXISTS_WITH_SAME_NAME];
```

```
          }
        }
        await MongOrb('Client').collection.updateOne(
          { _id: src._id },
          {
            $set: {
              ...args.client,
            },
          },
        );
      },
```

It is important for us to verify whether a user intends to change their email address to a value that already exists in our database. We just perform a check to see whether the email address or phone number was already used by somebody else. If everything is okay with those parameters, we can fire an `updateOne` function.

The next resolver is `salonClientOps`:

```
salonClientOps: async (yoga, args) => {
    const src = yoga[0] as ClientModel;
    return MongOrb('SalonClient').collection.findOne({
      _id: args._id,
      client: src._id,
    });
  },
```

Here, we have a simple pipe that returns the `SalonClient` object responsible for the connection between the `Client` and `SalonProfile` types.

The last resolver is responsible for connecting the client and the salon, and that is `registerToSalonResolver`:

```
registerToSalon: async (yoga, args) => {
  const src = yoga[0] as ClientModel;
  const Salon = await
    MongOrb('Salon').collection.findOne({
      slug: args.salonSlug,
    });
  if (!Salon) throw new GraphQLError('Salon with the
  following slug does not exist');

  const exists = await
  MongOrb('SalonClient').collection.findOne({
    salon: Salon._id,
```

```
          client: src._id,
        });
        if (exists) {
          throw new GraphQLError('This client already exist
          in this salon');
        }
        const result = await
          MongOrb('SalonClient').createWithAutoFields(
            '_id',
            'updatedAt',
            'createdAt',
          )({
            client: src._id,
            salon: Salon._id,
            visits: [],
          });
        return !!result.insertedId;
      },
```

Here, we begin by looking to see whether the salon with slug provided by the schema consumer already exists. We decided that we will require slug (a human-readable, unique identifier) instead of _id because it is easier to remember for the salon client. Then, we check whether the connection is already made and, if it is not, we proceed with creating the SalonClient object.

SalonClientOps

In SalonClientOps, we need to register a visit and send a message to the salon. Here is the code for src/SalonClientOps.ts:

```
import { createResolvers } from '@/src/axolotl.js';
import { MongOrb, SalonClientModel } from '@/src/orm.js';
import {
  VisitError,
  VisitStatus
} from '@/src/zeus/index.js';
import { GraphQLError } from 'graphql';

export default createResolvers({
  SalonClientOps: {
// resolver code goes here
  },
});
```

First, let's take a look at the `createVisit` resolver. Its role is to create a `Visit` object with the service requested by the salon client:

```
createVisit: async (yoga, args) => {
  const src = yoga[0] as SalonClientModel;
  try {
    new Date(args.visit.whenDateTime);
  } catch (error) {
    return [VisitError.INVALID_DATE];
  }
  await MongOrb('Visit').createWithAutoFields(
    '_id',
    'createdAt',
    'updatedAt',
  )({
    client: src.client,
    service: args.visit.serviceId,
    whenDateTime: args.visit.whenDateTime,
    status: VisitStatus.CREATED,
  });
  return;
},
```

Here, we check that the date is valid, and if it isn't, we return the invalid date error as an enum. If it is correct, we can create the `Visit` object from its arguments and source.

The second resolver is the `sendMessage` resolver, used to communicate between the salon and the client, and is almost identical to the one provided by `SalonOps`. Here is what it looks like:

```
sendMessage: async (yoga, args) => {
  const src = yoga[0] as SalonClientModel;
  const thread = await MongOrb('MessageThread')
    .collection.findOneAndUpdate(
      {
        salonClient: src._id,
      },
      {
        $setOnInsert: {
          salonClient: src._id,
        },
      },
      {
        upsert: true,
      },
```

```
    );
    if (!thread) throw new GraphQLError('Corrupted
      message thread. Please try again');
    const result = await
      MongOrb('Message').createWithAutoFields(
        '_id',
        'createdAt',
        'updatedAt',
      )({
        messageThread: thread._id,
        sender: src._id,
        message: args.message.message,
      });
    return !!result.insertedId;
  },
```

Thanks to the `SalonClient` object, we have only one connection between `messageThread` and `SalonClient`. It means that there is only one big conversation between a client and a salon.

Wrapping up the resolvers

Now, we need to join all the resolvers we created together. To do this, we need to connect them inside the main `src/resolvers.ts` file, like so:

```
import AuthPayload from '@/src/AuthPayload.js';
import Client from '@/src/Client.js';
import ClientOps from '@/src/ClientOps.js';
import ClientQuery from '@/src/ClientQuery.js';
import Message from '@/src/Message.js';
import MessageThread from '@/src/MessageThread.js';
import Mutation from '@/src/Mutation.js';
import PublicMutation from '@/src/PublicMutation.js';
import Query from '@/src/Query.js';
import SalonAnalytics from '@/src/SalonAnalytics.js';
import SalonClient from '@/src/SalonClient.js';
import SalonClientOps from '@/src/SalonClientOps.js';
import SalonOps from '@/src/SalonOps.js';
import SalonProfile from '@/src/SalonProfile.js';
import SalonQuery from '@/src/SalonQuery.js';
import Service from '@/src/Service.js';
import ServiceOps from '@/src/ServiceOps.js';
import UserOps from '@/src/UserOps.js';
import UserQuery from '@/src/UserQuery.js';
import Visit from '@/src/Visit.js';
```

```
import VisitOps from '@/src/VisitOps.js';
import { createResolvers } from '@/src/axolotl.js';

const resolvers = createResolvers({
    ...Query,
    ...Mutation,
    ...AuthPayload,
    ...ClientQuery,
    ...ClientOps,
    ...Client,
    ...MessageThread,
    ...Message,
    ...PublicMutation,
    ...SalonProfile,
    ...SalonAnalytics,
    ...SalonOps,
    ...SalonClientOps,
    ...SalonClient,
    ...SalonQuery,
    ...Service,
    ...ServiceOps,
    ...UserQuery,
    ...UserOps,
    ...Visit,
    ...VisitOps,
});

export default resolvers;
```

Here, we imported all the resolvers from separate files and used our `createResolvers` function to join them, without losing type safety.

Then, we can import those resolvers inside the `src/index.ts` file:

```
import resolvers from '@/src/resolvers.js';
import {
  graphqlYogaAdapter
} from '@aexol/axolotl-graphql-yoga';
const p = process.env.PORT || '4000';
graphqlYogaAdapter(resolvers).listen(parseInt(p), () => {
  console.log('LISTENING to ' + p);
});
```

That's it. Now, run the following command to start the development server:

```
npm run dev
```

Then, inside another **Terminal** tab, run this command to start the local MongoDB:

```
npm run mongo-dev
```

Everything should work successfully. I hope you like my pattern of creating GraphQL servers.

Now, let's take a look at how we can write some tests to check that everything works properly.

Writing integration tests

To test our backend, we will write integration tests for our GraphQL server. With these test files, we will check whether the user can log in, register, check user data, add a salon, and add a service to the salon.

These tests validate the interactions between the GraphQL server, its resolvers, and the underlying data sources. They verify that the GraphQL schema, queries, mutations, and subscriptions are functioning correctly, as well as any associated business logic. More widely, integration tests enable you to catch potential issues in the integration layer, ensuring the overall reliability and stability of the GraphQL backend.

For this purpose, we will be utilizing **Vitest**, which is a cutting-edge testing framework powered by Vite. It offers seamless code transformation and testing capabilities, even with TypeScript and ES Modules. The setup process for Vitest is considerably simpler compared to Jest, making it an ideal choice for our testing needs.

Setting up the tests

Before we start writing the tests, we need to set up the test environment for Vitest with TypeScript and ES Modules. To do this, we need a separate tsconfig file for building our app, as we don't want to include test code in the transpiled output. So, create a file named tsconfig.build.json and add this code:

```
{
  "extends":"./tsconfig.json",
  "exclude": [
    "lib",
    "node_modules",
    "vitest.config.ts",
    "**/*.test.ts"
  ]
}
```

Here, we added a line excluding files ending with test.ts in all folders.

Then, install `vitest` and the required plugins:

```
npm i -D vitest vite-tsconfig-paths unplugin-swc @swc/core
```

Next, to make `vitest` work with our path setup and TypeScript, we have to create a `./vitest.config.ts` file with the following code:

```
import path from 'path';
import { defineConfig } from 'vitest/config';
import swc from 'unplugin-swc';
import tsconfigPaths from 'vite-tsconfig-paths';

export default defineConfig({
  test: {
    include: ['src/**/*.test.ts'],
    disableConsoleIntercept: true,
  },
  plugins: [swc.vite(), tsconfigPaths()],
});
```

Now, we can add a line in the `package.json` file's `scripts` property and change the `build` and `watch` commands to use the new `tsconfig`:

```
"scripts": {
  "test": "vitest",
  "build": "tspc -p tsconfig.build.json",
  "watch": "tspc -p tsconfig.build.json --watch",
},
```

Finally, inside the `nodemon.json` file, add this line to its `ignore` field:

```
"src/**/*.test.ts"
```

We do this because we don't want to recompile the app when the test file changes.

Okay, we have now finished the setup and can move on to writing actual tests inside our project.

Writing the test files

All the test files should be located in the `src` folder and have the extension `.test.ts`.

Our first test will be used to test the `PublicMutation.ts` resolvers and will be named `PublicMutation.test.ts`. Inside the created file, add the following content:

```
import { Chain } from '@/src/zeus/index.js';
import { expect, test } from 'vitest';
```

```
const HOST = 'http://localhost:4000/graphql';

const client = Chain(HOST);
```

Here, we imported `Chain` to be used as a schema consumer-facing, type-safe GraphQL client. We will use it to write type-safe queries and mutations. We also imported `expect` and `test` from `vitest` – the `test` function is utilized to define a single test within our testing framework, while the `expect` function serves as a standard testing function, where it expects certain circumstances to be fulfilled or met during the test execution.

Next, we need to actually implement `test`:

```
test('Should register a user and allow for the user to log
in', async () => {
  const username = Math.random().toString(8);
  const password = Math.random().toString(8);
  const registerResult = await client('mutation')({
    public: {
      register: [
        {
          username,
          password,
        },
        {
          token: true,
        },
      ],
    },
  });
  expect(registerResult.public?.register.token)
    .toBeTruthy();
  const loginResult = await client('mutation')({
    public: {
      login: [{ password, username }, { token: true }],
    },
  });
  expect(loginResult.public?.login?.token).toBeTruthy();
  const authorizedClient = Chain(HOST, {
    headers: { Authorization:
      `${loginResult.public?.login?.token}`,
        'Content-Type': 'application/json' },
  });
  const meResult = await authorizedClient('query')({
    user: {
```

```
      me: {
        username: true,
      },
    },
  });
  expect(meResult.user?.me.username).toEqual(username);
});
```

At the beginning, the test generates a random username and password and saves them to the `username` and `password` variables. Then, we use them to register a user and try to log in. If everything went well and both functions returned tokens, then we check whether we can fetch the username to make sure authorization is also working well.

After testing the authorization mechanism, inside our `UserOps` tests, we will check whether we can register a salon, register a client, and fetch client and salon data. The content of the `src/UserOps.test.ts` file is as follows:

```
import { Chain } from '@/src/zeus/index.js';
import { expect, test, beforeAll } from 'vitest';
const HOST = 'http://localhost:4000/graphql';

const user = {
  client: Chain(HOST),
};
beforeAll(async () => {
  const registerResult = await user.client('mutation')({
    public: {
      register: [
        {
          username: Math.random().toString(8),
          password: Math.random().toString(8),
        },
        {
          token: true,
        },
      ],
    },
  });
  user.client = Chain(HOST, {
    headers: { Authorization:
      `${registerResult.public?.register.token}`,
        'Content-Type': 'application/json' },
  });
});
```

Here, we are preparing the client for further queries inside the `beforeAll` function. Inside it, we register a user and create an authorized client instance inside our `user` variable's `client` value. This way, we don't need to authorize users in every test separately.

Next, we will write the test for salon registration in the file:

```
test('Should register a salon', async () => {
  const random = Math.random().toString(8);
  const [name, slug] = ['Dev hairdresser ' + random,
    'dev-hairdresser-' + random];
  const salonRegistration = await user.client('mutation')({
    user: {
      registerAsSalon: [
        {
          salon: {
            name,
            slug,
          },
        },
        {
          errors: true,
        },
      ],
    },
  });
  expect(salonRegistration.user?.registerAsSalon?.errors)
    .toBeFalsy();
  const salonService = await user.client('query')({
    user: {
      salon: {
        me: {
          name: true,
        },
      },
    },
  });
  expect(salonService.user?.salon?.me.name).toEqual(name);
});
```

To begin, we generate a random number and convert it into a string. Using this value, we create a name and slug by combining it with predefined names. Afterward, we execute the `registerAsSalon` mutation using our authorized client that was set up in the `beforeAll` block. We then proceed to verify that there are no errors and whether we are able to fetch the name of the salon successfully.

Next, let's write the test for the client:

```
test('Should register a client', async () => {
  const random = Math.random().toString(8);
  const phone = (100000000 + Math.floor(Math.random() *
    899999999)).toString();
  const [firstName, lastName, email] = ['John', 'Hubble',
    `${random}@example.com`];
  const clientRegistration = await
  user.client('mutation')({
    user: {
      registerAsClient: [
        {
          client: {
            firstName,
            lastName,
            email,
            phone,
          },
        },
        {
          errors: true,
        },
      ],
    },
  });
  expect(clientRegistration.user?.registerAsClient?.errors)
    .toBeFalsy();
  const clientService = await user.client('query')({
    user: {
      client: {
        me: {
          firstName: true,
          lastName: true,
          email: true,
          phone: true,
        },
      },
    },
  });
  expect(clientService.user?.client?.me.firstName)
    .toEqual(firstName);
  expect(clientService.user?.client?.me.lastName)
    .toEqual(lastName);
```

```
  expect(clientService.user?.client?.me.email)
    .toEqual(email);
  expect(clientService.user?.client?.me.phone)
    .toEqual(phone);
});
```

Here, we register a client and check all the client properties using its `ClientQuery.me` resolver.

To run all the tests, write the following in your terminal:

```
npm run test
```

Vitest will run in **Watch** mode, which means it will restart on every test file change. Here is the result you should get if everything is working properly:

```
√ src/UserOps.test.ts (2)
√ src/PublicMutation.test.ts (1)

Test Files  2 passed (2)
     Tests  3 passed (3)
  Start at  13:06:25
  Duration  221ms (transform 41ms, setup 0ms, collect
  110ms, tests 103ms, environment 0ms, prepare 77ms)
PASS  Waiting for file changes...
```

> **Note**
> Writing tests with Vitest is an effective testing method, but another option to consider is starting a database with a different name dedicated to tests and then killing it after the tests are done.

Summary

In this chapter, we started by creating a plan and a schema for our salon service. Then, we implemented the backend using known patterns with TypeScript, Node.js, MongoDB, GraphQL Yoga, GraphQL Zeus, and Axolotl. We also learned how to write integration tests for GraphQL backends using the Vitest testing tool.

In the next chapter, we will write frontend code for our salon app and finish this book with a full stack application.

15

From an Idea to a Working Project – Frontend Integration with GraphQL and TypeScript

In the previous chapter, we used GraphQL to build the backend for a small business management system that a hairdressers or beauty salon could use. In this chapter, we will delve into the exciting world of building a frontend development, and create a frontend for our system to bring it to life.

In this chapter, you will gain a solid understanding of how to create a frontend application using GraphQL, Vite, GraphQL Zeus, and shadcn. Our objective is to design a clean and aesthetically pleasing user interface, as well as to establish a user experience that facilitates seamless authentication with roles. Additionally, we will aim to create a GraphQL client and GraphQL layer that offers a smooth and enjoyable development experience.

So, in this chapter, we will cover the following topics:

- Setting up the project
- Preparing the GraphQL layer
- Creating our app's frontend

> **Note**
> Psst, as this chapter is about the frontend, it will be a long one. Prepare yourself!

Technical requirements

To complete this chapter, you will need the following:

- Terminal and Node.js installed

- A GraphQL IDE such as VSCode, Sublime Text, Notepad++, or GraphQL Editor

- A basic knowledge of TypeScript and React.js

You can access the code in this chapter in the book's GitHub repository: `https://github.com/ PacktPublishing/GraphQL-Best-Practices/tree/main/chapter-15`

Setting up the project

To set up our project, we will follow similar instructions to those we followed back in *Chapter 9* – we will use ReactJS with Vite to create the project, as well as Tailwind CSS together with the shadcn CLI for styling. So, let's get started:

1. Start by creating a project with Vite:

   ```
   npm create vite@latest
   ```

   When prompted, choose ReactJS and TypeScript within the interactive mode of this command. Then, run the following:

   ```
   cd PROJECT_NAME
   ```

2. Install `tailwind` and `postcss`:

   ```
   npm install -D tailwindcss postcss autoprefixer
   npx tailwindcss init -p
   ```

3. Add the following code to the `tsconfig.json` file, to load dependencies from absolute paths beginning with @:

   ```
   {
     "compilerOptions": {
       // ...
       "baseUrl": ".",
       "paths": {
         "@/*": [
           "./src/*"
         ]
       }
       // ...
     }
   }
   ```

4. Install the `vite-tsconfig-paths` plugin to resolve paths from `tsconfig.json`:

    ```
    npm i -D vite-tsconfig-paths
    ```

5. Set up the `vite.config.ts` file and include the `tsconfigPaths` plugin for path support:

    ```
    import { defineConfig } from 'vite'
    import react from '@vitejs/plugin-react'
    import tsconfigPaths from 'vite-tsconfig-paths';

    // https://vitejs.dev/config/
    export default defineConfig({
      plugins: [
        tsconfigPaths(),
        react(),
      ],
    });
    ```

6. Initiate shadcn, and for each question, set the following options:

    ```
    npx shadcn-ui@latest init
    ✔ Would you like to use TypeScript (recommended)? … yes
    ✔ Which style would you like to use? › Default
    ✔ Which color would you like to use as base color? › Stone
    ✔ Where is your global CSS file? … src/index.css
    ✔ Would you like to use CSS variables for colors? … no / yes
    ✔ Are you using a custom tailwind prefix eg. tw-? (Leave blank
    if not) …
    ✔ Where is your tailwind.config.js located? … tailwind.config.
    js
    ✔ Configure the import alias for components: … @/components
    ✔ Configure the import alias for utils: … @/utils
    ✔ Are you using React Server Components? … no / yes
    ✔ Write configuration to components.json. Proceed? … yes
    ```

7. Next, install GraphQL Zeus:

    ```
    npm i -D graphql-zeus
    ```

8. Add the zeus script to the `package.json` file `scripts` field:

    ```
    "zeus": "zeus http://localhost:4000/graphql ./src"
    ```

9. Run your backend from the previous chapter in a separate terminal (it should be running on port 4000). You can also run the backend from the previous chapter from the book's repository.

10. Then run the `npm zeus` command in a separate terminal:

```
npm run zeus
```

This should create a `src/zeus` repository with typings generated from your backend.

11. Install React Router, which will handle all the navigation in our project:

```
npm install react-router-dom
```

12. Install `jotai`, which we will use to manage the global state of the application:

```
npm i jotai
```

13. Install `slugify` to change strings into slugs:

```
npm i slugify
```

14. Install `date-fns` to efficiently work with `Date` objects:

```
npm i date-fns
```

15. Install `prettier` and the `eslint` plugin for it, which we will use to format our code:

```
npm i -D prettier eslint-config-prettier eslint-plugin-prettier
```

16. Add the `.prettierrc` file to the root folder with the following content:

```
{
    "trailingComma": "all",
    "tabWidth": 2,
    "semi": true,
    "singleQuote": true,
    "printWidth": 120,
    "bracketSpacing": true,
    "endOfLine": "lf",}
```

Prettier is responsible for ensuring the same print width and indentation across the file, making the code files look better.

17. Add the `prettier` plugin to your `.eslintrc.cjs` file. The full configuration content of `eslint` should look like this:

```
module.exports = {
  root: true,
  env: { browser: true, es2020: true },
  extends: [
    'eslint:recommended',
    'plugin:@typescript-eslint/recommended',
    "plugin:react/recommended",
```

```
  'plugin:react-hooks/recommended',
  "plugin:prettier/recommended" // Enables eslint-
  plugin-prettier and eslint-config-prettier. This
  will display prettier errors as ESLint errors.
  Make sure this is always the last configuration in
  the extends array.
],
ignorePatterns: ['dist', '.eslintrc.cjs'],
parser: '@typescript-eslint/parser',
plugins: ['react','react-refresh'],
rules: {
  'react-refresh/only-export-components': [
    'warn',
    { allowConstantExport: true },
  ],
  "react/react-in-jsx-scope":"off",
  "react-hooks/exhaustive-deps":"off"
},
}
```

Here, we used only the most important settings of `eslint`. We connected `eslint` to `prettier`, TypeScript, and React, and added React hooks dependency validation.

18. Finally, install all the required shadcn components:

```
npx shadcn-ui@latest add card button label input textarea dialog
```

19. Now, go inside the `src/main.tsx` file, which should look like this (if not, replace the content with the following code):

```
import React from "react";
import ReactDOM from "react-dom/client";
import App from "./App.tsx";
import "./index.css";

ReactDOM.createRoot(document.getElementById("root")!)
.render(
  <React.StrictMode>
    <App />
  </React.StrictMode>,
);
```

Within this component, we render the React application into the HTML `div` element with the `"root"` ID.

20. Finally, we import the `index.css` file generated by Tailwind CSS, which serves as the styling file for our application. Here is the `index.css` content:

```
@tailwind base;
@tailwind components;
@tailwind utilities;

#root,body,html{
    width: 100%;
    height: 100%;
}
```

At the top of this file, we can observe the Tailwind directives that import all the Tailwind features into our project. We also set the full width and height of our main components.

At this point, we have completed another lengthy setup. But hold on – this time, we need to prepare a GraphQL layer.

Preparing the GraphQL layer

Before we start creating components and styling them, we need to create a good communication layer between our backend and frontend. To do this, we will write selectors and prepare the client to communicate with the backend.

Here, we will expand upon the layer we started in *Chapter 9*. The reason behind this is we want to prepare a seamless way to communicate with the backend to limit the amount of communication code within React components to a minimum.

Writing selectors

Like in *Chapter 9*, we will create the `src/graphql/selectors.ts` file and then add content inside it to create the type-safe communication layer with our GraphQL types.

Let's start making the selectors. The first selectors will be for the client and salon, which we will use inside the hooks later:

```
import { FromSelector, Selector } from '@/zeus';
export const ClientSelector = Selector('Client')({
  _id: true,
  createdAt: true,
  firstName: true,
  lastName: true,
  email: true,
  phone: true,
});
```

```
export type ClientType =
  FromSelector<typeof ClientSelector, 'Client'>;
```

As a reminder, selectors are like GraphQL query fragments. Here, the client selector is responsible for fetching basic profile data. We also export the selector type – handy because we won't need to write types by hand later on.

Next, we need selectors to fetch the salons and match them to the registered client:

```
export const ServiceSelector = Selector('Service')({
  _id: true,
  name: true,
  price: true,
  description: true,
  createdAt: true,
  approximateDurationInMinutes: true,
});

export type ServiceType =
  FromSelector<typeof ServiceSelector, 'Service'>;

export const SalonProfileSelector =
  Selector('SalonProfile')
({
  _id: true,
  createdAt: true,
  name: true,
  slug: true,
  services: ServiceSelector,
});

export type SalonType = FromSelector<
  typeof SalonProfileSelector,
  'SalonProfile'
>;
```

Through `SalonProfileSelector`, we will fetch both the salon and its services in one call (that's why we included `ServiceSelector` inside the selector too).

Moving on, we need selectors that will be used by `SalonProfile` to select the `Visit` and `Client` fields connected to `SalonProfile`:

```
export const VisitSalonSelector = Selector('Visit')({
  _id: true,
```

```
  createdAt: true,
  whenDateTime: true,
  client: ClientSelector,
  status: true,
  service: ServiceSelector,
});

export const SalonClientForSalonSelector =
  Selector('SalonClient')
({
  _id: true,
  client: ClientSelector,
  createdAt: true,
});
```

With VisitSalonSelector, when fetching every visit, we are also fetching the client connected to the visit and the service that the client booked. Meanwhile, with SalonClientForSalonSelector, we fetch the _id field of the connection between the salon and the client and the actual client data.

We also need a selector for messaging purposes:

```
export const MessagesSelector = Selector('SalonClient')({
  messageThread: {
    messages: {
      message: true,
      createdAt: true,
      sender: {
        __typename: true,
        '...on SalonClient': {
          _id: true,
          client: ClientSelector,
        },
        '...on SalonProfile': {
          _id: true,
        },
      },
    },
  },
});
```

This selector needs some explanation. There is a single message thread between SalonProfile and Client. Within this selector, we retrieve the __typename field to distinguish messages from SalonProfile and Client.

In order to determine the sender of a message, we need to resolve a union. By examining the type name, we can determine whether the message originated from `SalonProfile` or the client's side. If the message is from `SalonProfile`, based on our pipe structure, we are aware that we, as the salon, are the owners of the message. On the other hand, if the message is from a client, we can verify the owner by obtaining the `_id` field, as we have already requested the message from a specific client. It is worth noting that we can reuse the same selector for client queries, streamlining the process.

Next, we will write the selector for the full `SalonQuery.me` operation to fetch all the salon data (without messages) in one query:

```
export const FullSalonMeQuerySelector =
  Selector('SalonQuery')
({
  me: SalonProfileSelector,
  visits: [
    { filterDates: { from: new Date().toISOString() } },
    VisitSalonSelector,
  ],
  clients: SalonClientForSalonSelector,
});

export type FullSalonMeQueryType = FromSelector<
  typeof FullSalonMeQuerySelector,
  'SalonQuery'
>;
```

Here, we mostly reused the existing selectors, using them like fragments in GraphQL to compose a selector for the full query.

Now, we will proceed to implement selectors for the client within the same file – these selectors will also be based on the `SalonClient` type, but this time they will be utilized by the client instead of `SalonProfile`:

```
const VisitClientSelector = Selector('Visit')({
  _id: true,
  createdAt: true,
  whenDateTime: true,
  service: ServiceSelector,
});
export const SalonClientListForClientSelector =
  Selector('SalonClient')
({
  _id: true,
  createdAt: true,
  salon: {
```

```
      name: true,
      _id: true,
      slug: true,
    },
  });

  export type SalonClientListForClientType = FromSelector<
    typeof SalonClientListForClientSelector,
    'SalonClient'
  >;

  export const SalonClientDetailForClientSelector =
    Selector('SalonClient')
  ({
    _id: true,
    createdAt: true,
    salon: {
      name: true,
      _id: true,
      slug: true,
      services: ServiceSelector,
    },
    visits: [
      { filterDates: { from: new Date().toISOString() } },
      VisitClientSelector,
    ],
  });

  export type SalonClientDetailForClientType = FromSelector<
    typeof SalonClientDetailForClientSelector,
    'SalonClient'
  >;
```

Inside `SalonClientListForClientSelector`, we get all the salons where the client registered with their basic data. The other selector, `SalonClientDetailForClientSelector`, will be used to fetch the individual salon details. Having two separate selectors for fetching the *list* of objects and the *details* of the object means we avoid over-fetching the data.

Implementing the client hook with authentication and authorization

To execute all the queries with selectors, we need a simple client hook to hold our authorization data.

> **Note**
>
> The code in this section will be similar to that implemented in *Chapter 9*. If you can, try to complete the section without further instructions.

As a quick reminder, React hooks allow developers to use state and other React features in functional components. They provide a way to manage component state and lifecycle methods without the need for class components. Hooks enable easier code reuse and logic extraction and make it simpler to write clean and concise code in React applications.

We will put our main client hook in the `src/graphql/client.ts` file, side by side with our selector code. The `@/atoms` file will be specified below this file:

```ts
import { jwtToken } from "@/atoms";
import { Chain, HOST } from "@/zeus";
import { useAtom } from "jotai";
import { useMemo } from "react";

export const useClient = () => {
  const [token, setToken] = useAtom(jwtToken);
  const client = useMemo(() => {
    return Chain(HOST, {
      headers: {
        "Content-Type": "application/json",
        ...(token
          ? {
              Authorization: token,
            }
          : {}),
      },
    });
  }, [token]);
  const login = useCallback(
    (username: string, password: string) => {
      client('mutation')({
        public: {
          login: [
            { username, password },
            {
              token: true,
            },
          ],
        },
      }).then((response) => {
```

```
        const token = response.public?.login.token;
        if (token) {
          setToken(token);
        }
      });
    },
    [client, setToken],
  );

  const register = useCallback(
    (username: string, password: string) => {
      client('mutation')({
        public: {
          register: [
            { username, password },
            {
              token: true,
            },
          ],
        },
      }).then((response) => {
        const token = response.public?.register.token;
        if (token) {
          setToken(token);
        }
      });
    },
    [client, setToken],
  );

  return {
    client,
    setToken,
    login,
    register,
    isLoggedIn: !!token,
  };
};
```

Breaking down all of that code, first of all, we import jwtToken from an atom (as another reminder, an **atom** is responsible for holding a variable's value and re-rendering the component using the atom every time the value of the atom changes). Here, we are referring to a jotai atom from the src/atoms.ts file, which looks like this:

```
import { atom } from "jotai";

export const jwtToken = atom("token", "");
```

You can also refer to the example code in the repository, where I have implemented persistence using local storage for this atom. This feature is crucial as it ensures that the user remains logged in even if they close the browser tab or window.

Next, we import the Chain function, which is a type-safe fetch wrapper for GraphQL backends, and a HOST constant, which is the same host we provided in the zeus script inside the package.json file.

Inside the useClient hook, we consume the jwtToken atom from jotai to allow reading and setting a value on it – so all the components that use it react to the changes. Then, we memoize the client based on that token. Before the token is set – meaning before the user is logged in, we return the unauthorized client, which, after the user logs in, is transformed into the authorized client.

We implemented login and register functions to call our backend mutations and store tokens after successful login/register operations.

In the end, we return the client, setToken, login, and register functions, as well as the isLoggedIn helper variable, from our React hook.

Implementing dedicated hooks

Now, before delving into the realm of React components, we will implement dedicated GraphQL hooks that contain the necessary queries. Having dedicated GraphQL hooks is beneficial as it helps prevent bloating the component code by separating the logic for fetching data from the rendering logic. This separation allows for better organization and maintainability of code within the application.

Hooks for registered users

When you register as a user of our service, you will still need to create a client and/or salon account to use the app. Let's create the useMeQueries hook inside src/pages/me/useMeQueries.ts with the following content:

```
import { useClient } from '@/graphql/client';
import { ResolverInputTypes } from '@/zeus';
import { useCallback } from 'react';
import { useNavigate } from 'react-router-dom';
```

```
export const useMeQueries = () => {
  const { client } = useClient();
  const n = useNavigate();

  const registerAsSalon = useCallback(
    (salon: ResolverInputTypes['CreateSalon']) => {
      return client('mutation')({
        user: {
          registerAsSalon: [
            {
              salon,
            },
            {
              errors: true,
            },
          ],
        },
      }).then((r) => {
        if (!r.user?.registerAsSalon?.errors) {
          n('/me/salon');
        }
      });
    },
    [client, n],
  );
  const registerAsClient = useCallback(
    (clientForm: ResolverInputTypes['CreateClient']) => {
      return client('mutation')({
        user: {
          registerAsClient: [
            {
              client: clientForm,
            },
            {
              errors: true,
            },
          ],
        },
      }).then((r) => {
        if (!r.user?.registerAsClient?.errors) {
          n('/me/client');
        }
      });
```

```
    },
    [client, n],
  );
  return { registerAsSalon, registerAsClient };
};
```

Inside this hook, we have two functions – the first function is used to register as a salon, and the second is to register as a client. In the functions, we used a helper type, `ResolverInputTypes`, to get the input parameters for those queries.

If registering is successful, we will be redirected to the home of the client or salon. If it is not, we will be redirected to the user's home page.

Hooks for salons

As we can now register as a salon, we need to prepare hooks for the salon owner. Those hooks should allow the salon owner to edit the data of the salon. These dedicated GraphQL hooks will also interact with services and visits, enabling the creation and modification of these entities.

Let's start by defining the imports and state variables inside our hook:

```
import { useClient } from '@/graphql/client';
import {
  FullSalonMeQuerySelector,
  FullSalonMeQueryType,
} from '@/graphql/selectors';
import { useCallback, useState } from 'react';
import { format, subDays } from 'date-fns';
import { ResolverInputTypes } from '@/zeus';

export const useSalonQueries = () => {
  const { client } = useClient();
  const [mySalon, setMySalon] =
    useState<FullSalonMeQueryType>();
```

At the beginning of the function, we used our client hook and declared the `mySalon` state variable. We then used the exported `FullSalonMeQueryType` to hold all the salon data in one state variable.

To set this data inside the `mySalon` state variable, we will use the `fetchMe` and `me` functions:

```
  const me = useCallback(() => {
    return client('query')({
      user: {
        salon: FullSalonMeQuerySelector,
      },
    }).then((r) => r.user?.salon);
```

```
  }, [client]);

  const fetchMe = useCallback(() => {
    me().then((result) => {
      setMySalon(result);
    });
  }, [me]);
```

The me function executes the query to get an owned salon, while the fetchMe function stops setMySalon from being exposed outside of the hook. Every time we need to refetch the data from the backend, we will call this function from outside the hook to set the mySalon state variable.

Now, we will write queries to manipulate our Service objects:

```
  const createService = useCallback(
    (service: ResolverInputTypes['CreateService']) => {
      return client('mutation')({
        user: {
          salon: {
            createService: [{ service }, true],
          },
        },
      }).then((r) => {
        fetchMe();
        return r.user?.salon?.createService;
      });
    },
    [client, fetchMe],
  );

  const updateService = useCallback(
    (_id: string, service:
      ResolverInputTypes['UpdateService']
    ) => {
      return client('mutation')({
        user: {
          salon: {
            serviceOps: [
              { _id },
              {
                update: [{ service }, true],
              },
            ],
          },
```

```
      },
    }).then((r) => {
      fetchMe();
      return r.user?.salon?.serviceOps?.update;
    });
  },
  [client, fetchMe],
);


const deleteService = useCallback(
  (_id: string) => {
    client('mutation')({
      user: {
        salon: {
          serviceOps: [
            { _id },
            {
              delete: true,
            },
          ],
        },
      },
    }).then(() => fetchMe());
  },
  [client, fetchMe],
);
```

For the `createService`, `updateService`, and `deleteService` queries, we use `ResolverInputTypes` to pass the parameters, and after every mutation, we use the `fetchMe` function to refetch the data. Those functions are used to trigger the mutations responsible for adding a service, updating it, and deleting it.

You can also add resolvers to manipulate `visits` using the same pattern, like so:

```
const createVisit = useCallback((visit:
  ResolverInputTypes['CreateVisitFromAdmin']) => {
    return client('mutation')({
      user: {
        salon: {
          createVisit: [{ visit }, true],
        },
      },
    }).then((r) => {
      fetchMe();
```

```
        return r.user?.salon?.createVisit;
      });
    },
    [client, fetchMe],
  );
  const updateVisit = useCallback(
    (_id: string, visit:
      ResolverInputTypes['UpdateVisitFromAdmin']
    ) => {
      return client('mutation')({
        user: {
          salon: {
            visitOps: [
              { _id },
              {
                update: [{ visit }, { errors: true }],
              },
            ],
          },
        },
      }).then((r) => {
        const errors =
          r.user?.salon?.visitOps?.update?.errors;
        if (!errors) fetchMe();
        return errors;
      });
    },
    [client, fetchMe],
  );
  const deleteVisit = useCallback(
    (_id: string) => {
      return client('mutation')({
        user: {
          salon: {
            visitOps: [
              { _id },
              {
                delete: true,
              },
            ],
          },
        },
      }).then((r) => {
```

```
        fetchMe();
        return r.user?.salon?.visitOps?.delete;
      });
    },
    [client, fetchMe],
  );
```

We added functions that manage the `Visit` objects. Those functions trigger mutations responsible for creating, updating, and deleting the `Visit` objects. They also refetch the salon data after successful completion.

We will also have a function to fetch salon analytics:

```
const analytics = useCallback(() => {
  return client('query')({
    user: {
      salon: {
        analytics: [
          {
            filterDates: {
              from: format(subDays(new Date(), 90),
              'YYYY-MM-DD'),
            },
          },
          {
            cashPerDay: {
              amount: true,
              date: true,
            },
            visitsPerDay: {
              amount: true,
              date: true,
            },
          },
        ],
      },
    },
  }).then((r) => r.user?.salon);
}, [client]);
```

Our `analytics` function fetches the data for the last 90 days. To do that, we used `date-fns`, `subDays`, and `format` functions to find the date 90 days ago from now. Then, we requested `visitsPerDay` and `cashPerDay` to be returned by the resolver.

The last function will be used for messaging purposes:

```
const messages = useCallback(
  (salonClientId: string) => {
    return client('query')({
      user: {
        salon: {
          client: [{ _id: salonClientId },
          MessagesSelector],
        },
      },
    }).then((r) => r.user?.salon?.client?.messageThread);
  },
  [client],
);

const sendMessage = useCallback(
  (salonClientId: string, message: string) => {
    return client('mutation')({
      user: {
        salon: {
          sendMessage: [{ salonClientId,
          message: { message } }, true],
        },
      },
    });
  },
  [client],
);
```

For the `messages` function, we need to provide `salonClientId`. This will fetch all the messages between `SalonProfile` and `Client`.

To send a message, we also need `salonClientId` and the message's content. Then, we pass it to our `sendMessage` mutation to send the message.

To finish the salon hook, we need to return all the functions at the end of it:

```
return {
  fetchMe,
  mySalon,
  analytics,
  createService,
  updateService,
  deleteService,
```

```
      createVisit,
      updateVisit,
      deleteVisit,
      messages,
      sendMessage,
    };
};
```

With the salon hook done, we can move on and implement queries for the client.

Hooks for the salon client

As a salon client, we need fewer queries – we just want to see the salon, the services we booked, and the visits. We will start with the imports and implementing the state variable:

```
import { useClient } from '@/graphql/client';
import {
  ClientSelector,
  MessagesSelector,
  SalonClientListForClientSelector,
  SalonClientListForClientType,
} from '@/graphql/selectors';
import { ResolverInputTypes } from '@/zeus';
import { useCallback, useEffect, useState } from 'react';

export const useClientQueries = () => {
  const { client } = useClient();
  const [salonClients, setSalonClients] = useState<
    SalonClientListForClientType[]
  >([]);
```

Inside our state variable, we will hold all the connected salons with their basic data.

To fetch that data and put it inside our `salonClients` variable, we will use the following functions. First, we have the `getSalonClients` function:

```
  const getSalonClients = useCallback(() => {
    return client('query')({
      user: {
        client: {
          clients: SalonClientListForClientSelector,
        },
      },
    }).then((r) => {
      setSalonClients(r.user?.client?.clients || []);
```

```
        return;
    });
}, [client]);

useEffect(() => {
    getSalonClients();
}, [getSalonClients]);
```

In our `getSalonClients` function, we fetch the data of all the salons connected to the client and then use the result of the function to set the `salonClients` variable. Then, we create a `useEffect` hook to trigger that function when the hook appears inside the code.

Next, we implement the `salonClientById` function to get the individual salon:

```
const salonClientById = useCallback(
    (_id: string) => {
        return client('query')({
            user: {
                client: {
                    client: [{ _id },
                    SalonClientDetailForClientSelector],
                },
            },
        });
    },
    [client],
);
```

With this function, we fetch the details of the salon, such as the services and visits booked at that salon. Instead of setting the result of the query inside the `salonClientById` function to a state variable, we will return it. This is because the function will be reused for different salons that the client is connected to.

But wait, before we can fetch any salons as a client, we need to register a client to the salon:

```
const registerToSalon = useCallback(
    (slug: string) => {
        return client('mutation')({
            user: {
                client: {
                    registerToSalon: [{ salonSlug: slug }, true],
                },
            },
        }).then((r) => {
            if (r.user?.client?.registerToSalon) {
```

```
          getSalonClients();
          return true;
        }
      });
    },
    [client],
  );
```

Inside the `registerToSalon` function, we call the `registerToSalon` mutation with the slug provided by the user. It needs to reflect the real-world situation that a hairdresser gives us this `slug` value and it is easy to write on a piece of paper.

Now let's move on to the functions responsible for client data:

```
const me = useCallback(() => {
  return client('query')({
    user: {
      client: {
        me: ClientSelector,
      },
    },
  }).then((r) => r.user?.client?.me);
}, [client]);
const update = useCallback(
  (clientUpdate: ResolverInputTypes['UpdateClient']) => {
    return client('mutation')({
      user: {
        client: {
          update: [
            { client: clientUpdate },
            {
              errors: true,
            },
          ],
        },
      },
    }).then((r) => r.user?.client?.update?.errors);
  },
  [client],
);
```

Here, we have the me function to return the client's personal data, such as their first name, email, and so on, and the `update` function allows the client to update this personal data.

Now, as we have most of the functions implemented in our hook, we can implement our main client functionality – booking visits at the salon. Here is the function that wraps the `createVisit` resolver:

```
const createVisit = useCallback((salonId: string, visit:
  ResolverInputTypes['CreateVisitFromClient']) => {
    return client('mutation')({
      user: {
        client: {
          salonClientOps: [
            { _id: salonId },
            {
              createVisit: [{ visit }, { errors: true }],
            },
          ],
        },
      },
    }).then((r) =>
      r.user?.client?.salonClientOps?.createVisit?.errors
    );
  },
  [client],
);
```

This function creates a `Visit` object for a salon client. If an operation or process encounters an error or failure, it will return error messages or indications of the failure.

The only functions left – `messages` and `sendMessage` – are responsible for sending messages to the salon:

```
const messages = useCallback(
  (salonClientId: string) => {
    return client('query')({
      user: {
        client: {
          client: [{ _id: salonClientId },
          MessagesSelector],
        },
      },
    }).then((r) =>
      r.user?.client?.client?.messageThread);
  },
  [client],
);
```

```
const sendMessage = useCallback((salonId: string,
  message: ResolverInputTypes['MessageInput']) => {
    return client('mutation')({
      user: {
        client: {
          salonClientOps: [
            { _id: salonId },
            {
              sendMessage: [{ message }, true],
            },
          ],
        },
      },
    }).then((r) =>
      r.user?.client?.salonClientOps?.sendMessage);
  },
  [client],
);
```

To finish the salon client hook, we need to return all the functions at the end of the hook:

```
return {
  getSalonClients,
  salonClients,
  salonClientById,
  update,
  createVisit,
  registerToSalon,
  messages,
  sendMessage,
  me,
};
```

While these functions look similar to the functions we already created inside our `useSalonQueries` hook, here, we have utilized the client pipe, which informs our backend we are a client, not a salon. This is important because it helps us to do the following:

- Avoid over-fetching data

- Avoid optional fields dependent on the fetcher leading to less code

- Separate the logic of the client and the salon, leading to better code maintainability

Now that we have our hooks in place, we need to implement additional atoms to hold data for the salon and client.

Creating base atoms

We will hold the client and salon data inside dedicated jotai atoms – this way, we can fetch the client and salon data inside guard components and consume their data in view components.

So, add the following content to the src/atoms.ts file:

```
import { ClientType } from '@/graphql/selectors';
import { atom } from 'jotai';

const atomWithLocalStorage = <T>(key: string,
initialValue: T) => {
  const getInitialValue = () => {
    const item = localStorage.getItem(key);
    if (item !== null) {
      return JSON.parse(item) as T;
    }
    return initialValue;
  };
  const baseAtom = atom(getInitialValue());
  const derivedAtom = atom(
    (get) => get(baseAtom),
    (get, set, update) => {
      const nextValue =
        typeof update === 'function' ?
          update(get(baseAtom)) : update;
      set(baseAtom, nextValue);
      localStorage.setItem(key, JSON.stringify(nextValue));
    },
  );
  return derivedAtom;
};

export const jwtToken =
  atomWithLocalStorage('loyolo-token', '');
export const clientData =
  atom<ClientType | undefined>(undefined);
```

We started by defining the function to persist atoms inside localStorage for the token. Then, we added two plain atoms:

- clientData to store personal client data
- salonData to store the salon data

With the atoms in place, we can move on and implement the role guards needed by the project.

Setting up multiple guards

In the *Chapter 9* project, we had the `AuthGuard` component responsible for checking whether users were logged in and redirecting them to the login page if they were not.

Inside this project, we need to create more guard components – after the user is logged in inside both the client and salon parts of the app, we need to check whether they have registered as a salon or client and redirect them to the registration form if they are not.

To do this, we need three guard components – `AuthGuard`, `ClientGuard`, and `SalonGuard`.

AuthGuard

`AuthGuard` will have the same code as we had in *Chapter 9*, so add the same code inside the `src/pages/me/AuthGuard.tsx` file.

ClientGuard

Next, the `ClientGuard` component will be used to check whether the user has its own client object. If it has, it will render the children; if not, it will redirect them to the client registration page. Here is the code for the `src/pages/me/Client/ClientGuard.tsx` file:

```
import { clientData } from '@/atoms';
import { useClient } from '@/graphql/client';
import { ClientSelector } from '@/graphql/selectors';
import { useAtom } from 'jotai';
import { useEffect, useState } from 'react';
import {
  Outlet,
  useLocation,
  useNavigate
} from 'react-router-dom';
const ClientGuard = () => {
  const [, setClientData] = useAtom(clientData);
  const nav = useNavigate();
  const location = useLocation();
  const { client } = useClient();
  const [loading, setLoading] = useState(true);

  useEffect(() => {
    client('query')({
      user: {
        client: {
          me: ClientSelector,
        },
```

```
      },
    })
      .then((r) => {
        if (r.user?.client?.me._id) {
          setLoading(false);
          setClientData(r.user.client.me);
          return;
        }
        nav('/me/registerClient?' +
          `next=${location.pathname}`);
      })
      .catch(() => {
        nav('/me/registerClient?' +
          `next=${location.pathname}`);
      });
  }, [nav, location.pathname, client, setClientData]);

  return (
    <div
      className=
        "w-full h-full flex flex-col container space-y-8
        p-16"
    >
      {!loading && <Outlet />}
    </div>
  );
};
export default ClientGuard;
```

In this guard component, we have one useEffect hook and react-router-dom Outlet (we discussed the latter back in *Chapter 9*). We call the backend to fetch the client from the logged-in user. If there is no client object associated with the logged-in user, we redirect the user to the /me/registerClient page, which contains the client registration form. If the client exists, we set the client data to our clientData atom, leave them on the current page, and set loading to true to allow the rendering of the outlet.

SalonGuard

The SalonGuard component will do exactly the same thing that the client guard does but for the salon. We will check whether the user has the SalonProfile object associated with the GraphQL backend. Here is the code for src/pages/me/SalonGuard.tsx:

```
import { useClient } from '@/graphql/client';
import { SalonProfileSelector } from '@/graphql/selectors';
```

```
import { useEffect, useState } from 'react';
import {
  Outlet,
  useLocation,
  useNavigate
} from 'react-router-dom';
const SalonGuard = () => {
  const nav = useNavigate();
  const location = useLocation();
  const { client } = useClient();
  const [loading, setLoading] = useState(true);

  useEffect(() => {
    client('query')({
      user: {
        salon: {
          me: SalonProfileSelector,
        },
      },
    })
      .then((r) => {
        if (r.user?.salon?.me._id) {
          setLoading(false);
          return;
        }
        nav('/me/registerSalon?' +
          `next=${location.pathname}`);
      })
      .catch(() => {
        nav('/me/registerSalon?' +
          `next=${location.pathname}`);
      });
  }, [nav, location.pathname, client]);

  return (
    <div
      className=
        "w-full flex flex-col container space-y-8"
    >
      {!loading && <Outlet />}
    </div>
  );
};
export default SalonGuard;
```

The flow is the same as in the `ClientGuard` component, but we redirect the client to the salon registration form.

The biggest benefit of these guard components is that we don't have to check whether the user is authorized to see a certain page. The guard components check that and redirect the user to an appropriate registration/sign-up page if the user is not authorized. Without them, we would have to write custom hooks that would need to run on every page.

After all that work preparing the GraphQL layer, we are well-prepared to create the visual part of our frontend!

Creating our app's frontend

The frontend of our app will allow users to log in and register a user account. They should also be able to fill in registration forms for the salon or client. To do this, we will start by implementing the routing system and creating a layout for our app. After that, we will implement a home page, as well as the salon and client registration forms. At the end, we will code salon and client functionalities. So, let's get started!

Routing

Before we create the visual part of the app, we have to set up basic routing to see how the components behave while we develop the project.

To do this, inside `src/App.tsx` (which we imported inside `src/main.tsx`), we will create a routing tree for `react router` and return the `react router` app:

```
import {
  createBrowserRouter,
  RouterProvider
} from 'react-router-dom';
import Home from '@/pages/Home/Home';
import AuthRoot from '@/pages/auth/AuthRoot';
import Login from '@/pages/auth/Login';
import Register from '@/pages/auth/Register';
import Root from '@/Root';
import AuthGuard from '@/pages/me/AuthGuard';
import ClientGuard from '@/pages/me/Client/ClientGuard';
import SalonGuard from '@/pages/me/Salon/SalonGuard';
import RegisterClient from '@/pages/me/RegisterClient';
import RegisterSalon from '@/pages/me/RegisterSalon';
import ClientHome from '@/pages/me/Client/ClientHome';
import SalonHome from '@/pages/me/Salon/SalonHome';
```

```
const router = createBrowserRouter([
  {
    path: '',
    element: <Root />,
    children: [
      {
        path: '',
        element: <Home />,
      },
      {
        path: 'auth',
        element: <AuthRoot />,
        children: [
          {
            path: 'login',
            element: <Login />,
          },
          {
            path: 'sign-up',
            element: <Register />,
          },
        ],
      },
      {
        path: 'me',
        element: <AuthGuard />,
        children: [
          {
            path: 'client',
            element: <ClientGuard />,
            children: [
              {
                path: '',
                element: <ClientHome />,
              },
            ],
          },
          {
            path: 'salon',
            element: <SalonGuard />,
            children: [
              {
                path: '',
```

```
                element: <SalonHome />,
            },
        ],
      },
      {
        path: 'registerClient',
        element: <RegisterClient />,
      },
      {
        path: 'registerSalon',
        element: <RegisterSalon />,
      },
    ],
  },
  ],
  },
]);

function App() {
  return <RouterProvider router={router} />;
}

export default App;
```

> **Note**
>
> We will be making these pages later on, so you can come back here after you make every component to verify the imports of all the page components.

In *Chapter 9*, we learned how react-router-dom works. However, in our current implementation, we don't have the components necessary for the final form of our routing tree. One notable difference between our old routing tree and the updated version is the addition of two new guards – one for clients and another for salons, giving us more access levels.

Now let's move on and implement the registration and login forms.

Authentication and user authorization

In this project, we will have the same components for login and registration as we had in *Chapter 9*, so just copy the whole src/pages/auth folder from *Chapter 9* to src/pages/auth in this chapter.

After copying what we have already done, we will create a layout for our application.

Root layout

Our root layout will hold the information on whether the user is logged in. It will also wrap the page content inside a container so all the pages will have the same navigation bar. We will also allow users to navigate to the home page and login and registration forms.

Let's look at the code of our layout that we will put in `src/Root.tsx`:

```tsx
import { jwtToken } from '@/atoms';
import { Button } from '@/components/ui/button';
import { useAtom } from 'jotai';
import { Link, Outlet } from 'react-router-dom';

const Root = () => {
  const [token, setToken] = useAtom(jwtToken);
  return (
    <>
      <nav
        className=
          "w-full absolute hidden flex-col gap-6 text-lg
          font-medium md:flex md:flex-row md:items-center
          md:gap-5 md:text-sm lg:gap-6"
      >
        <div
          className="container flex items-center w-full">
          <Link
            to="/"
            className=
              "text-muted-foreground transition-colors
              hover:text-foreground px-8 py-4 block"
          >
            Home
          </Link>
          <div
            className=
              "ml-auto space-x-2 flex items-center"
          >
            {!!token && (
              <>
                <p>User logged in</p>
                <Button
                  variant={'outline'}
                  onClick={() => {
                    setToken(null);
```

```
              }}
            >
              Logout
            </Button>
          </>
        )}
        {!token && (
          <>
            <Link to="/auth/login">
              <Button>Login</Button>
            </Link>
            <Link to="/auth/sign-up">
              <Button variant={'outline'}>
                Register
              </Button>
            </Link>
          </>
        )}
      </div>
    </div>
  </nav>
  <div
    className=
      "flex w-full gap-4 md:ml-auto md:gap-2 lg:gap-4
      min-h-full pt-8 container"
  >
    <Outlet />
  </div>
  </>
  );
};
export default Root;
```

We started by defining the nav component. Inside it, we hold the information that states that if the user isn't logged in, we won't display anything, but if they are, we will display **Login** and **Register** buttons. Then, we have a div element with Tailwind CSS styling that contains the children of the layout.

After our layout is ready to hold all of our pages, we can actually implement the **Home** page.

Crafting the Home page

Our service is only for logged-in users, but we can create a home page that at least has some basic information and lets you choose your path. Our home page should look like *Figure 15.1*, where there are two paths: **I am a Client** and **I own a Salon**:

Home   Login   Register

Select your path

I am a Client   I own a Salon

Figure 15.1: Home page rendered

To create this, enter the following code in the `src/pages/Home/Home.tsx` file:

```
import { Button } from '@/components/ui/button';
import { Link } from 'react-router-dom';

const Home = () => {
  return (
    <div
      className=
        "flex flex-col space-y-16 container items-center
        pb-24"
    >
      <div
        className=
          "space-y-16 flex flex-col w-full items-center
          pt-[20vh]"
      >
        <h1
          className=
            "scroll-m-20 text-4xl font-extrabold
            tracking-tight lg:text-5xl"
        >
          Select your path
        </h1>
        <div
          className=
            "flex w-full max-w-md items-center
            justify-center space-x-4"
        >
```

```
                    <Link to="/me/client">
                      <Button className="px-16" type="submit">
                        I am a Client
                      </Button>
                    </Link>
                    <Link to="/me/salon">
                      <Button
                      className=
                        "px-16" variant="outline" type="submit"
                      >
                        I own a Salon
                      </Button>
                    </Link>
                </div>
              </div>
            </div>
          );
        };
        export default Home;
```

For the **Home** page, we render a header and the two buttons that lead to the part of our service that needs the user to be logged in. After clicking one of the buttons, if the user is not logged in, `AuthGuard` should stop them and display the login form. If the user is logged in but doesn't have the associated client or salon account, it should redirect them to the registration form.

Let's go and implement those registration forms.

Registration forms

We need to create two forms – one is the salon registration form and the other is the client registration form. We will work on the salon form first.

Salons are essential for initiating client-side operations, as clients are required to register with salons in order to book appointments. Therefore, we will begin by focusing on salon owners.

As we already did a lot in our GraphQL layer, our salon registration form is just a simple visual implementation connected to the layer. Here is the code you need to enter into `src/pages/me/RegisterSalon.tsx`:

```
import { Button } from '@/components/ui/button';
import {
  Card,
  CardContent,
  CardDescription,
  CardHeader,
```

```
    CardTitle,
  } from '@/components/ui/card';
import { Input } from '@/components/ui/input';
import { Label } from '@/components/ui/label';
import { useMeQueries } from '@/pages/me/useMeQueries';
import { ResolverInputTypes } from '@/zeus';
import { useEffect, useState } from 'react';
import slugify from 'slugify';

const RegisterSalon = () => {
  const [salonFormValues, setSalonFormValues] = useState<
    Partial<ResolverInputTypes['CreateSalon']>
  >({});
  const { registerAsSalon } = useMeQueries();
  useEffect(() => {
    setSalonFormValues((s) => ({
      ...s,
      slug: s.name ? slugify(s.name, { lower: true,
        trim: true }) : s.slug,
    }));
  }, [salonFormValues.name]);
```

We started the `RegisterSalon` component by importing the needed UI components – including `Card`, `Input`, and `Label` – and the `useMeQueries` hook that has the `register` functions inside.

Then, inside the component, we specified the `salonFormValues` state variable based on partial `ResolverInputTypes['CreateSalon']`. We used the `Partial` generic type because our form will be empty at the beginning and fields are required in the `CreateSalon` input.

After that, we extracted the `registerAsSalon` function from `useMeQueries` and implemented the `useEffect` hook that triggers the `slugify` function every time users input the name in the form. This way, we allow the salon to create its own slug, automatically creating the slug from the name that's input.

Nowhere is the JSX part responsible for displaying the form that submits the data to the `registerAsSalon` function:

```
  return (
    <div>
      <Card className="mx-auto max-w-sm">
        <CardHeader>
          <CardTitle className="text-xl">
            Create Salon
          </CardTitle>
          <CardDescription>
```

```
      Enter your information to create a salon
    </CardDescription>
  </CardHeader>
  <CardContent>
    <div className="grid gap-4">
      <Label htmlFor="first_name">Salon name</Label>
      <Input
        id="first_name"
        placeholder="John"
        required
        value={salonFormValues.name}
        onChange={(e) =>
          setSalonFormValues({
            ...salonFormValues,
            name: e.target.value,
          })
        }
      />
      <Label htmlFor="slug">My fancy salon</Label>
      <Input
        id="slug"
        placeholder="my-fancy-salon"
        required
        value={salonFormValues.slug}
        onChange={(e) =>
          setSalonFormValues({
            ...salonFormValues,
            slug: e.target.value,
          })
        }
      />
      <Button
        type="submit"
        className="w-full"
        onClick={() => {
          if (!salonFormValues.name ||
            !salonFormValues.slug) return;
          registerAsSalon({
            name: salonFormValues.name,
            slug: salonFormValues.slug,
          });
        }}
      >
```

```
            Create a salon account
          </Button>
        </div>
      </CardContent>
    </Card>
  </div>
);
};

export default RegisterSalon;
```

Here, we just render `Input` and `Label` components that are connected to our state variable. In the end, we have a button, **Create a salon account**. When the user clicks the button, it fires the `registerAsSalon` function using the form state variable.

Here is the result:

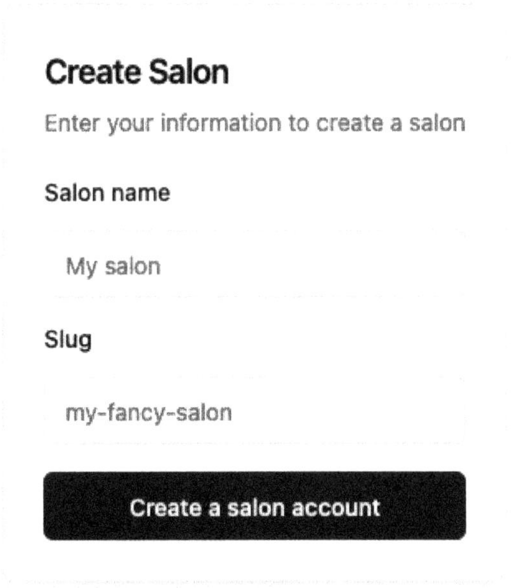

Create Salon

Enter your information to create a salon

Salon name

My salon

Slug

my-fancy-salon

Create a salon account

Figure 15.2: Salon registration form

As mentioned, the second form we need to implement is the client registration form. It will have more data and longer component code. To start, add this code to `src/pages/me/RegisterClient.tsx`:

```
import { Button } from '@/components/ui/button';
import {
  Card,
```

```
    CardContent,
    CardDescription,
    CardHeader,
    CardTitle,
  } from '@/components/ui/card';
import { Input } from '@/components/ui/input';
import { Label } from '@/components/ui/label';
import { useMeQueries } from '@/pages/me/useMeQueries';
import { ResolverInputTypes } from '@/zeus';
import { useState } from 'react';

const RegisterClient = () => {
  const [clientFormValues, setClientFormValues] = useState<
    Partial<ResolverInputTypes['CreateClient']>
  >({});
  const { registerAsClient } = useMeQueries();
```

After importing all the necessary functions, we specified the `clientFormValues` state variable and extracted the `registerAsClient` function from `useMeQueries`.

Next, we want the component JSX code to appear clean and well-organized, thanks to the separation of the logic portion. Here is the code for that:

```
  return (
    <div>
      <Card className="mx-auto max-w-sm">
        <CardHeader>
          <CardTitle className="text-xl">
            Sign Up
          </CardTitle>
          <CardDescription>
            Enter your information to create an account
          </CardDescription>
        </CardHeader>
        <CardContent>
          <div className="grid gap-4">
            <Label htmlFor="first_name">First name</Label>
            <Input
              id="first_name"
              placeholder="John"
              required
              value={clientFormValues.firstName}
              onChange={(e) =>
                setClientFormValues({
```

```
                    ...clientFormValues,
                    firstName: e.target.value,
                })
            }
        />
```

We can repeat the `Label` and `Input` parts for the remaining properties: `lastName`, `phone`, and `email`. Just copy the `Label` and `Input` content and change the `id`, `placeholder`, `value`, and `onChange` values.

> **Note**
>
> Please see the code in the book's GitHub repository to check whether you have done it correctly.

In the end, we have a button to create a client account:

```
            <Button
                type="submit"
                className="w-full"
                onClick={() => {
                    if (
                        !clientFormValues.firstName ||
                        !clientFormValues.lastName
                    )
                        return;
                    registerAsClient(
                        clientFormValues as
                        ResolverInputTypes['CreateClient'],
                    );
                }}
            >
                Create a client account
            </Button>
        </div>
      </CardContent>
    </Card>
  </div>
 );
};

export default RegisterClient;
```

Here, we implemented a button that fires a `registerAsClient` function. We cast the `clientFormValues` type to the `CreateClient` type because we already know that the form contains validated values and its response fits that type definition.

Here is the result of the second registration form:

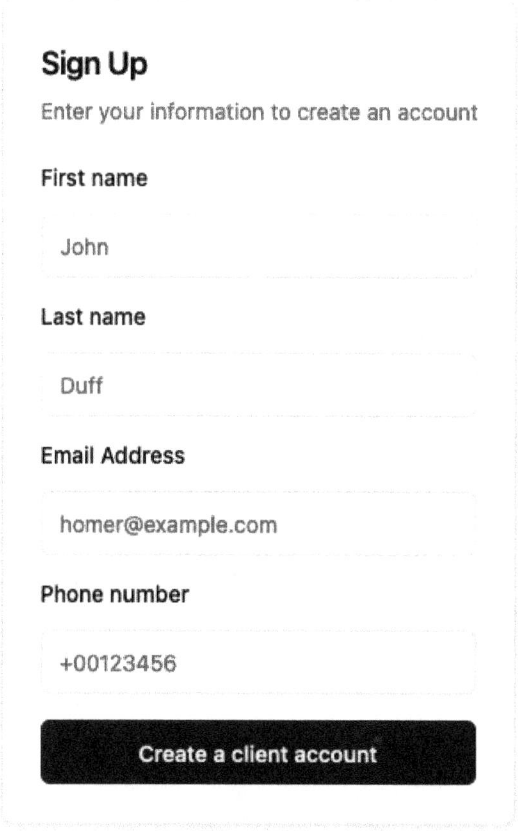

Sign Up

Enter your information to create an account

First name

John

Last name

Duff

Email Address

homer@example.com

Phone number

+00123456

Create a client account

Figure 15.3: Client registration form

Now, we can move on to the vital part of our project and implement salon owner functionalities.

Salon area

After the user registers their salon, they can automatically enter the page where they can manage their salon. In the salon area, we will code the following functions:

- The creation of a service
- The deletion of a service

- Displaying services in a list

- Displaying visits in a list

- Displaying clients in a list

To begin, add the following code to `src/pages/me/Salon/SalonHome.tsx`:

```
import { Button } from '@/components/ui/button';
import {
  Card,
  CardContent,
  CardDescription,
  CardFooter,
  CardHeader,
  CardTitle,
} from '@/components/ui/card';
import
  CreateServiceDialog
from '@/pages/me/Salon/CreateService';
import {
  useSalonQueries
} from '@/pages/me/Salon/useSalonQueries';
import { Trash } from 'lucide-react';
import { useEffect } from 'react';

const SalonHome = () => {
  const {
    fetchMe,
    mySalon,
    deleteService,
    createService
  } = useSalonQueries();

  useEffect(() => {
    fetchMe();
  }, []);
```

We start by specifying `mySalon` – this data is available here as we stored it in a hook code. Then, we use the `fetchMe` function to fetch more information about our salon.

Next, we can return the JSX code:

```
return (
    <div className="flex flex-col items-start space-y-4">
      <h2 className="text-lg">
```

```
      Hello, {mySalon?.name} - your slug:{' '}
      <b>{mySalon?.slug}</b>
    </h2>
    <div className="flex space-x-4">
      <CreateServiceDialog createService={createService} />
    </div>
```

Here, we created a header containing the username and message. After that, we inserted the `CreateServiceDialog` component, which is a popup with a form to create a `Service` object.

Now we need to display all the booked visits in the JSX component:

```
<div className="flex flex-col w-full space-y-2">
  <h3>My visits</h3>
  <div className="grid grid-cols-3 w-full gap-4">
    {mySalon?.visits.map((v) => (
      <Card key={v._id}>
        <CardHeader>
          <CardTitle>
            {`${new Date(v.whenDateTime)
              .toLocaleString()} -
                ${v.client.firstName}
                ${v.client.lastName}`}
          </CardTitle>
          <CardDescription>
            {v.status}
          </CardDescription>
        </CardHeader>
        <CardContent>{v.service.name}</CardContent>
      </Card>
    ))}
  </div>
</div>
```

Here, we listed all the visits clients booked in our salon and converted the date and time to our local date and time.

Then, we will list all the salon services:

```
<div className="flex flex-col w-full space-y-2">
  <h3>My services</h3>
  <div className="grid grid-cols-3 w-full gap-4">
    {mySalon?.me?.services?.map((s) => (
      <Card key={s._id}>
        <CardHeader>
```

```
          <CardTitle>{s.name}</CardTitle>
          <CardDescription>{s.price}</CardDescription>
        </CardHeader>
        <CardContent>{s.description}</CardContent>
        <CardFooter>
          <Button
            variant="destructive"
            onClick={() => {
              deleteService(s._id);
            }}
          >
            <Trash className="mr-2 h-4 w-4" />
            Delete
          </Button>
        </CardFooter>
      </Card>
    ))}
  </div>
</div>
```

Again, using the `Card` component, we listed all of the salon services with the option to delete them using the `deleteService` function.

Finally, the last part of the code holds the client list:

```
<div className="flex flex-col w-full space-y-2">
  <h3>My clients</h3>
  <div className="grid grid-cols-3 w-full gap-4">
    {mySalon?.clients.map((v) => (
      <Card key={v._id}>
        <CardHeader>
          <CardTitle>
            {`${v.client.firstName}
              ${v.client.lastName}`}
          </CardTitle>
          <CardDescription>
            {`Joined ${new Date(v.createdAt)
              .toLocaleString()}`}
          </CardDescription>
        </CardHeader>
      </Card>
    ))}
  </div>
</div>
```

```
    </div>
  );
};

export default SalonHome;
```

With the `SalonHome` component, we have developed visual components and incorporated logic from the GraphQL layer. The finished page should look like this:

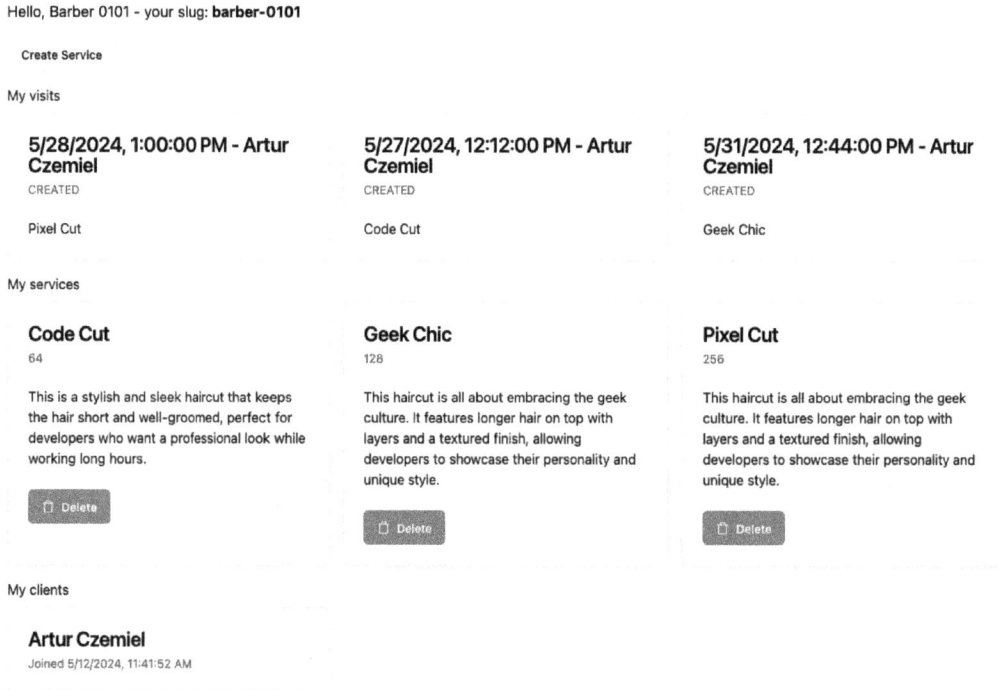

Hello, Barber 0101 - your slug: **barber-0101**

Create Service

My visits

5/28/2024, 1:00:00 PM - Artur Czemiel
CREATED

Pixel Cut

5/27/2024, 12:12:00 PM - Artur Czemiel
CREATED

Code Cut

5/31/2024, 12:44:00 PM - Artur Czemiel
CREATED

Geek Chic

My services

Code Cut
64

This is a stylish and sleek haircut that keeps the hair short and well-groomed, perfect for developers who want a professional look while working long hours.

🗑 Delete

Geek Chic
128

This haircut is all about embracing the geek culture. It features longer hair on top with layers and a textured finish, allowing developers to showcase their personality and unique style.

🗑 Delete

Pixel Cut
256

This haircut is all about embracing the geek culture. It features longer hair on top with layers and a textured finish, allowing developers to showcase their personality and unique style.

🗑 Delete

My clients

Artur Czemiel
Joined 5/12/2024, 11:41:52 AM

Figure 15.4: Salon panel

The only thing left to show you is the popup that will be used to create a `Service` object. It is made from simple visual components and dialogs from the UI library. Here is the code you need to put in `src/pages/me/Salon/CreateService.tsx`:

```
import { Button } from '@/components/ui/button';
import {
  DialogHeader,
  DialogFooter,
  Dialog,
  DialogContent,
  DialogDescription,
```

```
  DialogTitle,
  DialogTrigger,
} from '@/components/ui/dialog';
import { Input } from '@/components/ui/input';
import { Label } from '@/components/ui/label';
import { Textarea } from '@/components/ui/textarea';
import { ResolverInputTypes } from '@/zeus';

import { useEffect, useState } from 'react';

const CreateServiceDialog = ({
  createService,
}: {
  createService: (
    service: ResolverInputTypes['CreateService'],
  ) => Promise<string | undefined>;
}) => {
  const [dialogOpen, setDialogOpen] = useState(false);
  const [serviceForm, setServiceForm] =useState<
    Partial<ResolverInputTypes['CreateService']>
  >();
  useEffect(() => {
    setServiceForm({});
  }, [dialogOpen]);
```

Here, we imported all the components needed from our design library. Then we extracted the `createService` method from our `useSalonQueries` hook and declared the `dialogOpen` state variable to control the state of our dialog. After that, our `serviceForm` state variable will hold the user input from the form and the `useEffect` hook will clear the form every time the dialog is opened or closed.

Here is the JSX part of the component:

```
  return (
    <Dialog open={dialogOpen} onOpenChange={setDialogOpen}>
      <DialogTrigger asChild>
        <Button variant="outline">Create Service</Button>
      </DialogTrigger>
      <DialogContent className="sm:max-w-[425px]">
        <DialogHeader>
          <DialogTitle>Create Service</DialogTitle>
          <DialogDescription>
            Create Service you sell in your salon.
          </DialogDescription>
```

```
      </DialogHeader>
      <div className="grid gap-4 py-4">
        <div
          className=
            "grid grid-cols-4 items-center gap-4"
        >
          <Label htmlFor="name" className="text-right">
            Name
          </Label>
          <Input
            id="name"
            className="col-span-3"
            value={serviceForm?.name}
            onChange={(e) =>
              setServiceForm((s) =>
                ({ ...s, name: e.target.value }))
            }
          />
        </div>
        {/* more field components here */}
      </div>
```

Here, we have a `Label` component and an `Input` component to set the service name. Similar to what we did earlier, you can replicate the `Label` and `Input` sections for the following fields: `description`, `price`, and `approximateDurationInMinutes` (when working with the `price` and `duration` fields, please make sure to use the `"number"` input type property).

Finally, inside `DialogFooter`, we code the **Save changes** button together with the function to create a service:

```
      <DialogFooter>
        <Button
          type="submit"
          onClick={() => {
            if (
              serviceForm?.name &&
              serviceForm.price &&
              serviceForm.approximateDurationInMinutes &&
              serviceForm.description
            ) {
              createService(serviceForm as
              Required<typeof serviceForm>).then(
                (r) => {
                  if (r) {
```

```
                    setDialogOpen(false);
                  }
                },
              );
            }
          }}
        >
          Save changes
        </Button>
      </DialogFooter>
    </DialogContent>
  </Dialog>
  );
};
```

```
export default CreateServiceDialog;
```

This is the result of our code:

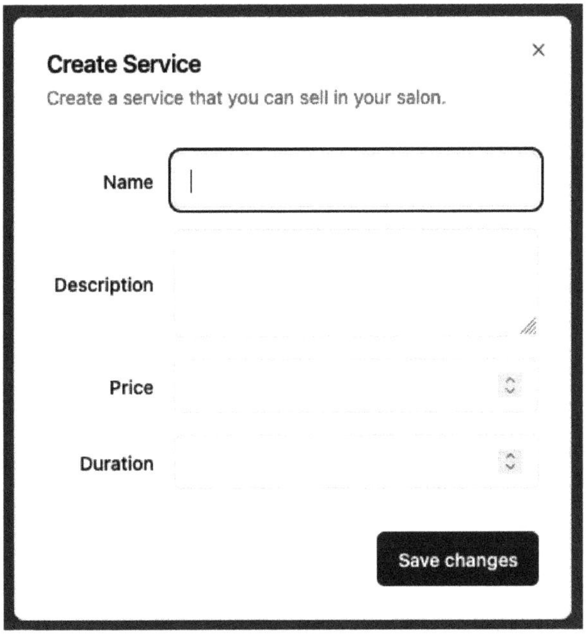

Figure 15.5: Create Service popup

When the service is successfully created, the dialog closes thanks to the `setDialogOpen(-false)` function.

After completing all the salon management parts, we need to enable our app for the salon clients.

Client area

On our client management page, we need to be able to register with a salon and book services in different salons. The client area is defined in the `src/pages/me/Client/ClientHome.tsx` file:

```
import { clientData } from '@/atoms';
import { Button } from '@/components/ui/button';
import {
  CardDescription,
  CardTitle,
  Card,
  CardHeader,
} from '@/components/ui/card';
import { Input } from '@/components/ui/input';
import { Label } from '@/components/ui/label';
import {
  SalonClientDetailForClientType
} from '@/graphql/selectors';
import ActiveSalon from '@/pages/me/Client/ActiveSalon';
import {
  useClientQueries
} from '@/pages/me/Client/useClientQueries';
import clsx from 'clsx';
import { useAtom } from 'jotai';
import { useEffect, useState } from 'react';

const ClientHome = () => {
  const [clientDataFromAtom] = useAtom(clientData);
  const {
    salonClients,
    registerToSalon,
    salonClientById
  } = useClientQueries();
  const [joinSalonSlug, setJoinSalonSlug] = useState('');
  const [activeSalonId, setActiveSalonId] = useState('');
  const [activeSalon, setActiveSalon] =
    useState<SalonClientDetailForClientType>();

  useEffect(() => {
    salonClientById(activeSalonId).then((s) =>
      setActiveSalon(s)
    );
    // eslint-disable-next-line react-hooks/exhaustive-deps
  }, [activeSalonId, setActiveSalon]);
```

We start by extracting all the needed functions from `useClientQueries`. Then, we declare two state variables:

- The first is `joinSalonSlug`, to hold the value of the input to join the salon.

- The second state variable is `activeSalonId`. We want clients to have the option to join different salons. When a client clicks on a salon tile, the `useEffect` function is triggered. This function fetches the detailed data of the selected salon and updates the `activeSalon` state variable, allowing the client to access information specific to that salon.

After displaying a header and relevant information, we present all the salons that the user has joined as individual cards:

```
return (
  <div className="flex flex-col space-y-8">
    <div className="flex flex-col space-y-2">
      <h2 className="font-bold text-xl">
        Hello, {clientDataFromAtom?.firstName}
        {clientDataFromAtom?.lastName}
      </h2>
      <p>
        Choose a salon to check your visits or book one
      </p>
      <div className="grid grid-cols-2 gap-4">
        {salonClients.map((sc) => (
          <Card
            className={clsx(
              'cursor-pointer hover:bg-gray-100',
              sc._id === activeSalonId && 'bg-gray-200',
            )}
            onClick={() => setActiveSalonId(sc._id)}
            key={sc._id}
          >
            <CardHeader>
              <CardTitle>{sc.salon.name}</CardTitle>
              <CardDescription>
                {`Joined on:${new Date(sc.createdAt)
                .toDateString()}`}
              </CardDescription>
            </CardHeader>
          </Card>
        ))}
      </div>
    </div>
```

```
        {activeSalon && (
          <ActiveSalon
            refetch={() =>
              salonClientById(activeSalonId).then((s) =>
                setActiveSalon(s)
              )
            }
            activeSalon={activeSalon}
          />
        )}
```

When a card is clicked and a salon is chosen, the `ActiveSalon` component is rendered on this page. The `refetch` function is passed as a prop to the `ActiveSalon` component, allowing it to refresh the `visits` data after a booking has been made.

Next, we just need the user to provide the salon slug and add a button to join the requested salon as their client:

```
      <div className="flex flex-col space-y-4">
          <h3>Join new salon</h3>
          <Label htmlFor="new-salon">Salon slug</Label>
          <Input
            id="new-salon"
            name="new-salon"
            value={joinSalonSlug}
            onChange={(e) =>
              setJoinSalonSlug(e.target.value)}
          />
          <Button
            onClick={() => {
              if (joinSalonSlug) {
                registerToSalon(joinSalonSlug);
              }
            }}
          >
            Join
          </Button>
        </div>
      </div>
    );
  };


export default ClientHome;
```

When the user clicks the **Join** button, the `registerToSalon` function is triggered with `joinSalonSlug` as a parameter.

Here is what the UI should look like:

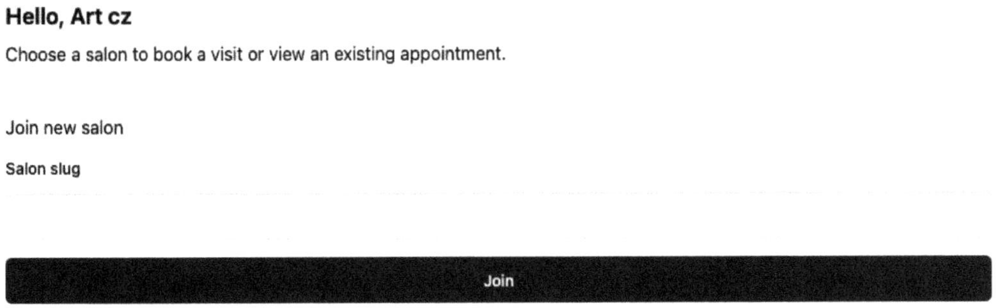

Hello, Art cz

Choose a salon to book a visit or view an existing appointment.

Join new salon

Salon slug

Join

Figure 15.6: Client home page

When the client clicks the salon card (in the previous figure, this would be **Barber0101**), the active salon page with all the salon details will appear. Its code is in `src/pages/me/Client/ActiveSalon.tsx`:

```
import { Button } from '@/components/ui/button';
import {
  CardDescription,
  CardTitle,
  Card,
  CardContent,
  CardFooter,
  CardHeader,
} from '@/components/ui/card';
import { Input } from '@/components/ui/input';
import {
  SalonClientDetailForClientType
} from '@/graphql/selectors';
import {
  useClientQueries
} from '@/pages/me/Client/useClientQueries';
import { useState } from 'react';

const ActiveSalon = ({
  activeSalon,
  refetch,
}: {
  activeSalon: SalonClientDetailForClientType;
```

```
  refetch: () => void;
}) => {
  const { createVisit} = useClientQueries();
  const [bookedVisitDateTime, setBookedVisitDateTime] =
    useState<string>();
```

Here, our well-prepared GraphQL layer allows us to easily specify the type of `activeSalon` parameter by importing `SalonClientDetailForClientType` from our GraphQL selectors and assigning it to the parameter. Then, we extracted the query needed to create a visit and create the `bookedVisitDateTime` state variable.

Next, we need to display a list of salon services and allow the user to set the date and time when they request a service:

```
return (
  <div className="flex flex-col space-y-4">
    <h5 className="font-bold text-lg">Your Visits</h5>
    {activeSalon.visits.map((v) => (
      <Card key={v._id}>
        <CardHeader>
          <CardTitle>
            {`${new Date(v.whenDateTime)
            .toLocaleString()} -
            ${v.service.name}`}
          </CardTitle>
          <CardDescription>
            {v.service.price}$
          </CardDescription>
        </CardHeader>
        <CardContent>
          {v.service.description}
        </CardContent>
      </Card>
    ))}
    <h4 className="font-bold text-lg">
      {activeSalon.salon.name} services
    </h4>
    <div className="grid grid-cols-2 gap-4">
      {activeSalon.salon.services?.map((sv) => (
        <Card key={sv._id}>
          <CardHeader>
            <CardTitle>{sv.name}</CardTitle>
            <CardDescription>
              {sv.price}$
```

```
              </CardDescription>
            </CardHeader>
            <CardContent>{sv.description}</CardContent>
            <CardFooter className="space-x-4">
              <Input
                type="datetime-local"
                value={bookedVisitDateTime}
                onChange={(e) => {
                  setBookedVisitDateTime(e.target.value);
                }}
              />
              <Button
                onClick={() => {
                  if (!bookedVisitDateTime) return;
                  createVisit(activeSalon._id, {
                    serviceId: sv._id,
                    whenDateTime: new Date(
                      bookedVisitDateTime
                    ).toISOString(),
                  }).then((r) => {
                    if (!r) {
                      refetch()
                    }
                  });
                }}
              >
                Book a visit
              </Button>
            </CardFooter>
          </Card>
        ))}
      </div>
    </div>
  );
};

export default ActiveSalon;
```

In this code, we display a list of all the visits and services available. Each service card includes a button and a date input, allowing users to book a visit for that particular service.

When the user clicks the **Book a visit** button, we format the date to an ISOString format and trigger the createVisit function.

Here is the detailed view of the salon client:

Your Visits

5/31/2024, 12:44:00 PM - Geek Chic
128$

This haircut is all about embracing the geek culture. It features longer hair on top with layers and a textured finish, allowing developers to showcase their personality and unique style.

Barber 0101 services

Code Cut
64$

This is a stylish and sleek haircut that keeps the hair short and well-groomed, perfect for developers who want a professional look while working long hours.

mm / dd / yyyy , -- : -- --   [ ]   Book a visit

Geek Chic
128$

This haircut is all about embracing the geek culture. It features longer hair on top with layers and a textured finish, allowing developers to showcase their personality and unique style.

mm / dd / yyyy , -- : -- --   [ ]   Book a visit

Figure 15.7: Detailed salon view

We did it! The frontend is done and you have finished the book!

Summary

In this chapter, we focused on preparing the GraphQL layer, putting everything responsible for communication with the backend in one layer. That made coding the visual part of our frontend really easy – we didn't have to think about this communication layer as it was well prepared.

Specifically, we learned how to work with guards and redirects and how to put data from the GraphQL backend inside type-safe state variables on the frontend. We also added another access level interconnected with the GraphQL layer.

At this point, our journey through the world of GraphQL comes to an end. In this book, I wanted to show you, from many angles, how important it is to have knowledge of GraphQL in today's world. We have gone through the specification, special design patterns, the integration of GraphQL with artificial intelligence, and finally, ended with full stack development.

We are already living in the era of artificial intelligence and mature development. You must remember that every project should start with a detailed description and specification, preferably also with a GraphQL schema.

And with these words, I would like to conclude this long journey. Thank you for making it all the way here. May GraphQL always be with you!

Index

‹packt›

www.packtpub.com

Subscribe to our online digital library for full access to over 7,000 books and videos, as well as industry leading tools to help you plan your personal development and advance your career. For more information, please visit our website.

Why subscribe?

- Spend less time learning and more time coding with practical eBooks and Videos from over 4,000 industry professionals

- Improve your learning with Skill Plans built especially for you

- Get a free eBook or video every month

- Fully searchable for easy access to vital information

- Copy and paste, print, and bookmark content

Did you know that Packt offers eBook versions of every book published, with PDF and ePub files available? You can upgrade to the eBook version at packtpub.com and as a print book customer, you are entitled to a discount on the eBook copy. Get in touch with us at customercare@packtpub.com for more details.

At www.packtpub.com, you can also read a collection of free technical articles, sign up for a range of free newsletters, and receive exclusive discounts and offers on Packt books and eBooks.

Other Books You May Enjoy

If you enjoyed this book, you may be interested in these other books by Packt:

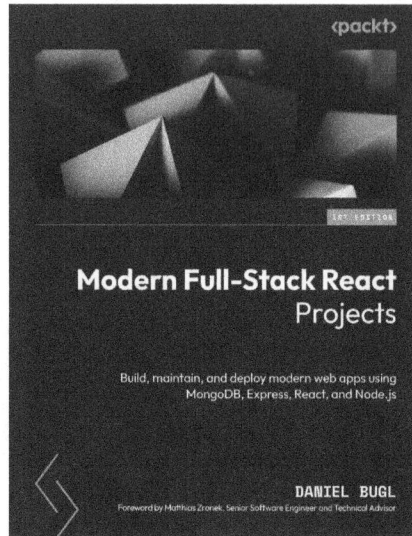

Modern Full-Stack React Projects

Daniel Bugl

ISBN: 978-1-83763-795-9

- Implement a backend using Express and MongoDB, and unit-test it with Jest
- Deploy full-stack web apps using Docker, set up CI/CD and end-to-end tests using Playwright
- Add authentication using JSON Web Tokens (JWT)
- Create a GraphQL backend and integrate it with a frontend using Apollo Client
- Build a chat app based on event-driven architecture using Socket.IO
- Facilitate Search Engine Optimization (SEO) and implement server-side rendering
- Use Next.js, an enterprise-ready full-stack framework, with React Server Components and Server Actions

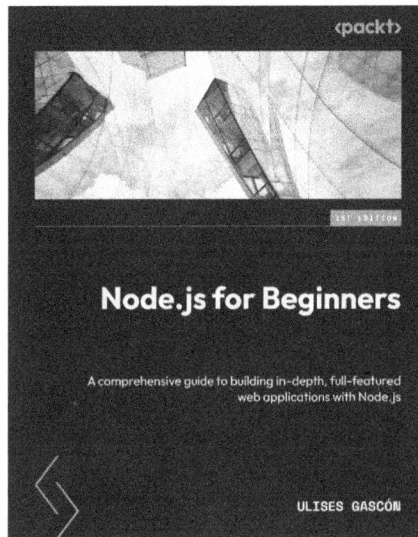

Node.js for Beginners

Ulises Gascón

ISBN: 978-1-80324-517-1

- Build solid and secure Node.js applications from scratch

- Discover how to consume and publish npm packages effectively

- Master patterns for refactoring and evolving your applications over time

- Gain a deep understanding of essential web development principles, including HTTP, RESTful API design, JWT, authentication, authorization, and error handling

- Implement robust testing strategies to enhance the quality and reliability of your applications

- Deploy your Node.js applications to production environments using Docker and PM2

Packt is searching for authors like you

If you're interested in becoming an author for Packt, please visit `authors.packtpub.com` and apply today. We have worked with thousands of developers and tech professionals, just like you, to help them share their insight with the global tech community. You can make a general application, apply for a specific hot topic that we are recruiting an author for, or submit your own idea.

Share Your Thoughts

Now you've finished *GraphQL Best Practices*, we'd love to hear your thoughts! Scan the QR code below to go straight to the Amazon review page for this book and share your feedback or leave a review on the site that you purchased it from.

`https://packt.link/r/1-835-46714-8`

Your review is important to us and the tech community and will help us make sure we're delivering excellent quality content.

Download a free PDF copy of this book

Thanks for purchasing this book!

Do you like to read on the go but are unable to carry your print books everywhere?

Is your eBook purchase not compatible with the device of your choice?

Don't worry, now with every Packt book you get a DRM-free PDF version of that book at no cost.

Read anywhere, any place, on any device. Search, copy, and paste code from your favorite technical books directly into your application.

The perks don't stop there, you can get exclusive access to discounts, newsletters, and great free content in your inbox daily

Follow these simple steps to get the benefits:

1. Scan the QR code or visit the link below

https://packt.link/free-ebook/978-1-83546-714-5

2. Submit your proof of purchase
3. That's it! We'll send your free PDF and other benefits to your email directly

www.ingramcontent.com/pod-product-compliance
Lightning Source LLC
Chambersburg PA
CBHW061742210326
41599CB00034B/6765